The Cytokine Network

Frontiers in Molecular Biology

SERIES EDITORS

B. D. Hames

Department of Biochemistry
and Molecular Biology
University of Leeds, Leeds LS2 9JT, UK

D. M. Glover

Cancer Research Laboratories,
Department of Anatomy and Physiology,
University of Dundee, Dundee DD1 4HN, UK

TITLES IN THE SERIES

The Cytokine Network

EDITED BY

Fran Balkwill

Biological Therapy Laboratory
Imperial Cancer Research Fund
Lincoln's Inn Fields
London WC2A 3PX

OXFORD

UNIVERSITY PRESS

OXFORD

UNIVERSITY PRESS

Great Clarendon Street, Oxford OX2 6DP

Oxford University Press is a department of the University of Oxford
and furthers the University's aim of excellence in research, scholarship,
and education by publishing worldwide in

Oxford New York

Athens Auckland Bangkok Bogotá Buenos Aires Calcutta
Cape Town Chennai Dar es Salaam Delhi Florence Hong Kong Istanbul
Karachi Kuala Lumpur Madrid Melbourne Mexico City Mumbai
Nairobi Paris São Paulo Singapore Taipei Tokyo Toronto Warsaw

with associated companies in Berlin Ibadan

Oxford is a registered trade mark of Oxford University Press
in the UK and in certain other countries

Published in the United States
by Oxford University Press Inc., New York

A catalogue record for this book is available from the British Library

Library of Congress Cataloging in Publication Data
(Data applied for)

ISBN 019 963 703 2 (hbk.)
ISBN 019 963 702 4 (pbk.)

Typeset by Footnote Graphics Warminster, Wilts

Printed in Great Britain by
The Bath Press, Avon

Preface

During the past ten years the phrase 'cytokine network' has become common currency in biomedical research. At first encompassing a few interleukins and variously named 'factors', the cytokine network now comprises several hundred proteins.

Cytokines contribute to a chemical signalling language in multicellular organisms that regulates development, tissue repair, haemopoiesis, inflammation, and the immune response. Potent cytokine polypeptides have pleiotropic activities and functional redundancy. They act in a complex network where one cytokine can influence the production of, and response to, many other cytokines. In the past five years, this bewildering array of effector molecules and associated cell-surface receptors has been simplified by assignment of cytokines and their receptors into structural superfamilies; elucidation of convergent intracellular signalling pathways; and molecular genetics, especially targeted gene disruption to 'knock-out' production of individual cytokines in mice. It is also now clear that the pathophysiology of infectious, autoimmune, and malignant disease can be partially explained by the induction of cytokines and the subsequent cellular response. Moreover, many viruses have evolved an ability to manipulate the cytokine network with homologues of cytokine ligands and receptors encoded in their genome, and functional polymorphisms of cytokine genes are now seen to contribute to the genetic programming of responses to pathogenic stimuli. Cloning of cytokine and cytokine receptor genes has allowed production of milligram quantities of purified protein for use in preclinical and clinical studies of acute and chronic infection, inflammatory disease, autoimmune disease, and cancer. Manipulation of the cytokine network with these recombinant proteins or other cytokine regulators provides a range of novel approaches to treating acute and chronic disease.

These current, exciting, and important themes are reflected in this book — dispatches from the frontiers of cytokine research.

London   F. B.
October, 1999

Contents

5 Chemokines   103

ALBERTO MANTOVANI AND SILVANO SOZZANI

8 Therapeutic manipulation of the cytokine network   174

FRAN BALKWILL

Contributors

FRAN BALKWILL
Imperial Cancer Research Fund, Lincoln's Inn Fields, London WC2A 3PX, UK.

FIONULA M. BRENNAN
Kennedy Institute of Rheumatology, 1 Aspenlea Road, Hammersmith, London W6 8LH, UK.

GORDON DUFF
Division of Molecular and Genetic Medicine, University of Sheffield, Royal Hallamshire Hospital, Sheffield S10 2JF, UK.

MICHAEL R. FANNON
Human Genome Sciences Inc., 9410 Key West Avenue, Rockville, Maryland 20850, USA.

MARC FELDMANN
Kennedy Institute of Rheumatology, 1 Aspenlea Road, Hammersmith, London W6 8LH, UK.

GIANNI GAROTTA
Human Genome Sciences Inc., 9410 Key West Avenue, Rockville, Maryland 20850, USA.

ALBERTO MANTOVANI
Istituto di Ricerche Farmacologiche 'Mario Negri', via Eritrea 62, 20157 Milano, Italy and Section of General Pathology, University of Brescia, 25123, Brescia, Italy.

KEATS NELMS
Laboratory of Immunology, NIAID, Bldg. 10/Room 11N323, National Institutes of Health, Bethesda, MD 20892–1892, USA.

SERGIO ROMAGNANI
Division of Clinical Immunology and Allergy, Institute of Internal Medicine and Immunoallergology, Instituto di Clinica Medica 111, University of Florence, 50134 Florence, Italy.

CRAIG A. ROSEN
Human Genome Sciences Inc., 9410 Key West Avenue, Rockville, Maryland 20850, USA.

STEVEN M. RUBEN
Human Genome Sciences Inc., 9410 Key West Avenue, Rockville, Maryland 20850, USA.

GEOFFREY L. SMITH
Sir William Dunn School of Pathology, University of Oxford, South Parks Road, Oxford OX1 3RE, UK.

SILVANO SOZZANI
Istituto di Ricerche Farmacologiche 'Mario Negri', via Eritrea 62, 20157 Milano, Italy.

JULIAN A. SYMONS
Sir William Dunn School of Pathology, University of Oxford, South Parks Road, Oxford OX1 3RE, UK.

Abbreviations

AIDS	acquired immune deficiency syndrome
APC	antigen-presenting cell
CIA	collagen-induced arthritis
CKR	cellular chemokine receptor
CRD	cysteine-rich domain
DC	dendritic cells
EBV	Epstein–Barr virus
EC	endothelial cell
EGF	epidermal growth factor
EMSA	electromobility shift assay
EST	expressed sequence tag
G-CSF	granulocyte colony-stimulating factor
GM-CSF	granulocyte-macrophage colony-stimulating factor
IDDM	insulin-dependent diabetes mellitus
IL	interleukin
IL-1ra	IL-1 receptor antagonist
IRS	insulin receptor substrate
JAK1	Janus kinase 1
KS	Kaposi's sarcoma
LIF	leukaemia inhibitory factor
LPS	lipopolysaccharide
MAPK	mitogen activated protein kinase
MCP	monocyte chemotactic protein
MDC	macrophage-derived chemokine
MHC	major histocompatibility complex
MIF	myeloid inhibitory factor
MIP	macrophage inflammatory protein
MPIF	myeloid progenitor inhibitory factor
MS	multiple sclerosis
MTD	maximum tolerated dose
NF-κB	nuclear factor-κB
NGF	nerve growth factor
ORF	open reading frame
PDGF	platelet-derived growth factor
PI	phosphoinositide
PTB	phosphotyrosine binding
PTK	protein tyrosine kinase
RA	rheumatoid arthritis

RFLP	restriction fragment length polymorphism
RT–PCR	reverse transcription polymerase chain reaction
SLE	systemic lupus erythematosis
SNP	single nucleotide polymorphism
Sos	Son Of Sevenless
SSc	systemic sclerosis
STAT	signal transducer and activator of transcription
TAM	tumour-associated macrophage
TCR	T cell receptor
TDT	transmission disequilibrium test
TGF	transforming growth factor
Th	T helper cells
Th1	Th Type 1
Th2	Th Type 2
TNF	tumour necrosis factor
TNFR	TNF receptor
VEGF	vascular endothelial growth factor
VGF	vaccinia growth factor
VNTR	variable number tandem repeats
VV	vaccinia virus

1 | A genomics approach to cytokine discovery

CRAIG A. ROSEN, MICHAEL R. FANNON, GIANNI GAROTTA,
AND STEVEN M. RUBEN

1. Introduction

With the human genome effort now well underway, the implications for gene discovery are becoming increasingly evident. As exemplified with cytokine discovery alone, we have witnessed significant expansion of gene families previously thought to be nearing completion and have identified completely new families. As the paradigm for discovery has shifted from identification of single genes to completion of gene families, so has the methodology used for characterization of gene function. In this respect, genomic approaches, such as differential gene display, biochip technology, and EST sequencing for gene expression and analysis, are all seeing increased use. Today, gene discovery of novel genes, using a genomics approach, can occur through two predominant routes. The first, which is a genetics approach (i.e. positional cloning), exploits the influence of a disease phenotype within families. The second approach is a sequence-based approach that involves high-throughput sequencing of thousands of cDNAs and compares sequence homology to pre-existing gene families and also examines tissue expression profiles to aid in prediction of gene function. This chapter will focus on the latter approach of gene discovery and will highlight how this approach has broadened our understanding of two cytokine families that include chemokines and tumour necrosis factor (TNF)-like proteins.

2. cDNA-based methods for high-throughput gene discovery

Our understanding of gene function and physiology at the molecular level is derived primarily from the identification and cloning of individual genes obtained from cDNA libraries. Through the use of diverse cDNA libraries, a wide variety of genes representative of many functionally distinct gene families has already been obtained. None the less, even with the tremendous progress over the last decade, the compendium of full-length and partial human genes, as represented in the full-length gene bank sequence repository GENBANK, probably covers less than 10% of expressed human genes.

In 1991, an approach designed to identify large numbers of expressed genes was developed. This approach, termed EST for 'expressed sequence tags', entails sequencing of short segments (150–600 bp) from either end of putative cDNAs (1, 2). A specific EST database, dbEST (3) has been established by the United States National Library of Medicine. If sequencing is performed on high quality cDNA libraries, the EST approach provides sequence information corresponding to the 5' and 3' ends of full-length genes. There now exist numerous high-throughput EST sequencing programs, both private and public. Although the sequencing approaches are fairly similar, the goals for individual projects differ considerably. Many of the academic efforts focus on identification of genes derived from specific tissues or physiological states (4–7). The Integrated Molecular Analysis of Genomes and their Expression (IMAGE) consortium prepares human and mouse cDNA libraries that are being sequenced by the Washington University Genome Center (St Louis, MO) under a grant from Merck and Company. The Expressed Gene Anatomy Database (EGAD), the human cDNA database initiated by The Institute for Genomic Research (TIGR; Gaithersburg, MD) collect and distribute information about human genes and gene expression. The Jackson Laboratories (Bar Harbor, ME), the MRC Human Genetics Unit (Edinburgh, UK) and University of Edinburgh (Edinburgh, UK), and several other laboratories are building a gene expression information resource for mouse development. The Cancer Genome Anatomy Project, CGAP (Bethesda, MD), is collecting gene expression information from analysis of normal and malignant tissues.

In the commercial sector, sequencing-enhanced gene identification efforts are focused on both identification of all expressed human genes and obtaining a complete understanding of gene expression profiles. To achieve this goal at Human Genome Sciences, cDNA libraries have been prepared from several hundred tissues. As genes expressed in one tissue are often absent in other tissues, representative libraries derived from both normal and activated states, as well as cell lines of normal and disease states, are included in the library set; libraries derived from embryonic and fetal tissues are also included because their gene expression patterns differ dramatically from that observed from adult tissue. With the standardization of tissue preparation, cDNA library construction and DNA template preparation, a meaningful comparison of gene expression profiles between libraries is possible. To ensure that the natural gene expression profile is maintained, the use of library normalization and subtraction technologies are minimized.

2.1 Data analysis

Bioinformatics efforts have generally focused on methods for evaluating sequence similarity and assessing the properties of related sets of sequences. Technology has been only recently developed to systematically estimate gene expression on a large scale. Carefully constructed EST databases can be viewed as a statistical sampling of genes being expressed in a variety of tissues, disease states, and developmental stages. Such databases enable us to estimate gene expression levels using computational methods. Thus, an EST database enables us to view gene expression in a new way. Rather than 'where is this gene expressed?', we can ask 'what genes are expressed in a specified tissue?' While the EST database lacks the resolving power of sophisticated expression assays, it offers a fast way to locate genes based on expression patterns. The database becomes a decision support tool that helps to identify candidates that will be more rigorously characterized in the laboratory.

2.1.1 Sequence analysis

Figure 1 identifies the types of analysis performed on new ESTs as they are sequenced. The analysis requires a database to be the repository for the sequence information, search results, library/tissue source, and related information. Rules that relate tissues, cDNA libraries, clones, and sequences are well understood by groups performing EST sequencing and straightforward to implement in a modern relational database management system.

With the random sampling approach of EST sequencing, multiple copies of the same cDNA are detected. This redundancy offers a way to estimate how often a gene is expressed in a given tissue or cell type. To work with this redundancy property, clusters of overlapping sequences are stored as a separate entity in the database (an *entity* is something the database stores information about). Sequence clusters provide

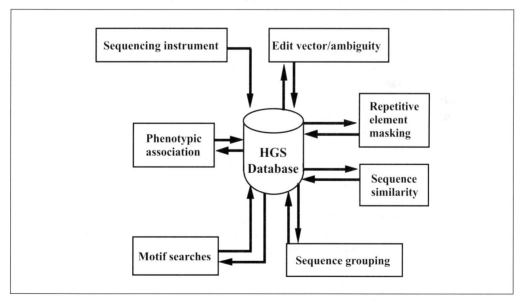

Fig. 1 Data collection and analysis. Every EST runs through a series of processing steps designed to (i) mask out vector, ambiguous, or repetitive sequences; (ii) determine if the EST is identical or similar to a known sequence; (iii) identify overlapping sequences in the database. Results of these processing steps are stored in an integrated database that may be queried by scientists looking for genes with specific characteristics.

a way to estimate which ESTs belong to the same cDNA; the correlation between clusters and cDNAs improves as overlapping fragments are added to the database.

The clustering software must allow for some ambiguities, insertions, deletions, and mismatches in alignments of overlapping sequences. If clustering parameters are too restrictive, the system fails to group sequences that belong together; if they allow too much ambiguity, the chance of false matches increases (8). Alternatively, spliced genes and genes with near-identical stretches of nucleotides within their cDNAs will be assigned to the same cluster if they fall within the selected clustering threshold. A minimum overlap of 50 nucleotides with at least 95% identity offers a good compromise for clustering parameters.

2.2   Selection method for candidate genes

To capitalize on the usefulness of the gene expression profile stored in the database, software must be implemented to identify and evaluate candidate genes. There are several expression questions we can pose to the database for insights that may lead to important genes.

(a) *What genes are expressed in a library or set of related libraries?* As a library is sequenced, the distribution of genes is a useful measure of library quality and productivity. Reviewing the list of abundantly expressed genes provides clues to the basic functions of a tissue or cell line. Running this analysis on sets of related libraries, such as all cancer libraries, highlights genes common to the set.

(b) *How does the expression pattern for one library compare with another?* Comparing the expression profile of normal and disease tissue may yield meaningful clues to disease mechanisms.

(c) *What genes are specific to a library or set of related libraries?* This powerful question can be answered if the database contains a very broad representation of tissues, disease states, and developmental stages. In this analysis, we look for genes that are highly specific to a library or set of libraries.

2.3   Case study of database analysis to identify a tissue-specific cytokine

We use an example from our work at Human Genome Sciences to illustrate how a new cytokine was discovered from analysis of ESTs stored in our database. The first step was to identify all the cDNA libraries from a tissue or cell type of interest. In this example, we selected the nine libraries derived from normal and diseased thymus. Next, we ran software to identify clusters where 70% or more of the ESTs were sequenced from the nine thymus libraries. The most abundant thymus-specific gene turned out to be a novel chemokine (Fig. 2A). Chemokines are a family of secreted cytokine molecules involved in immunoregulation and inflammatory processes mediated by their ability to attract leukocytes. Sequences from 17 independent clones

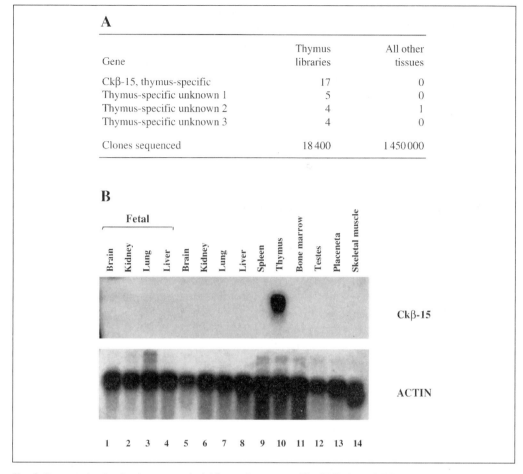

A

Gene	Thymus libraries	All other tissues
Ckβ-15, thymus-specific	17	0
Thymus-specific unknown 1	5	0
Thymus-specific unknown 2	4	1
Thymus-specific unknown 3	4	0
Clones sequenced	18 400	1 450 000

B

Fig. 2 Case study of a database search yielding a thymus-specific cDNA clone. HGS has sequenced ESTs from over 700 human cDNA libraries for which more than 1.5 million EST sequences have been derived. (A) The list shown depicts four clones identified as the most abundant thymus-specific genes derived from the sequencing project. (B) To confirm the accuracy of the computer-directed search results, a Northern blot analysis was performed using Ck-β15 as a probe.

were derived from two thymus libraries. The gene was not observed in random sampling of over 700 cDNA libraries and 1.4 million ESTs.

By assembling the ESTs into a consensus sequence, we can detect an open reading frame that is predicted to be a signal peptide by the *SignalP* program (9). *BLAST* analysis (10) of the ESTs confirm that the sequence is previously unreported and not similar to known human or non-human proteins. However, the sequence does contain a motif characteristic of chemokines (11, 12). Northern blot analysis (Fig. 2B) confirms the tissue specificity predicted from the database search. In subsequent analysis of the ESTs in the public domain, all matches were to sequences derived from thymus libraries.

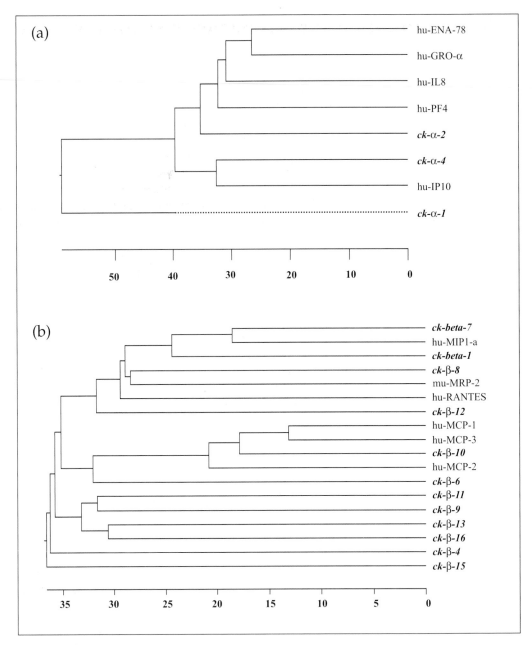

Fig. 3 (A) Phylogenetic trees of HGS α- and β-family chemokines. Alignment of chemokines by *Clustal* method using a PAM 250 matrix. (A) The phylogenetic tree shows the ancestral relationship among the reported α-chemokines. (B) The phylogenetic tree shows the ancestral relationship among the reported TNF-β chemokines. Novel chemokines identified by HGS are in bold italics. The length of each pair of branches represents the distance between sequence pairs, measured by number of non-conservative amino acid substitutions.

3. Identification of new chemokine family members

Chemokines are cytokines that elicit the chemotactic migration of leukocytes, stimulate pro-inflammatory activity, regulate angiogenesis, HIV viral entry and replication, and haematopoiesis (11, 12). Over the past several years, knowledge of the human/mouse chemokine family has witnessed considerable expansion (13). The chemokines exhibit 20–75% homology at the amino acid level and are classified into four families that are defined by cysteine signature motifs. The α-chemokines have the first two cysteines interrupted by an amino acid (CXC); the β-chemokines have the first two cysteines together (CC); the γ-chemokines contain only one N-terminal cysteine (C), and the fractalkines are membrane-bound proteins with the first two cysteines spaced by three additional amino acids (CX3C) (13). Searching for α- and β-chemokines motifs (CXC-[X]n-C-[X]m-C or CC-[X]n-C-[X]m-C where n = 22–24 and m =15–17).

HGS has identified and characterized three novel α-chemokines and 12 novel β-chemokines. As illustrated in Fig. 3, these newly discovered members have homologues across the entire chemokine family.

3.1 MPIF-1, a chemokine with stem cell inhibitory activity

Among the β-chemokines identified, myeloid progenitor inhibitory factor 1, MPIF-1 (also known as Ckβ-8 or MIP-5) is most homologous with the chemokine MIP-1α, displaying 51% identity and 67% similarity at the amino acid level (14). Like, MIP-1α, MPIF-1 mRNA is expressed at high levels in normal adult liver and lung. Lower levels of expression have been found in adult bone marrow, jejunum, ileum, colon, and fetal liver. Freshly isolated monocytes can be induced to express MPIF-1 messages following IL-1β or IFN-γ activation (15). Similar levels of message have also been detected in the myelomonocytic cell lines HL-60 and THP-1. At the protein level, MPIF-1 is secreted as a non-glycosylated polypeptide containing 99 amino acids with a molecular mass of 11.2 kDa. The *in vitro* analyses of MPIF-1 was based on 80 assays derived from normal and transformed cell populations and three major developmental cell lineages. The results from such experiments clearly indicate that MPIF-1 has no detectable effect on tumour cell lines usually used for screening of anti-cancer agents. Among normal cell types, the biological activity of MPIF-1 is restricted to specific cells within the peripheral immune system and the haematopoietic progenitor cell compartment (16). In particular, MPIF-1 has been found to:

- induce chemotactic responses in resting T cells and monocytes
- induce Ca^{2+} mobilization in monocytes, monocyte-derived dendritic cells, and eosinophils
- inhibit multipotential and committed progenitor colony formation
- inhibit human $CD34^+$ cell proliferation
- protect multipotential and committed progenitors from the cytotoxic effects of chemotherapeutic agents

In vivo analyses of MPIF-1 have been carried out in mice. The results from these studies indicate that MPIF-1 is a potent myeloprotectant for haematopoietic progenitor cells (17) (Fig. 4). The consequence of such protection is a more rapid recovery of bone marrow progenitors and peripheral cell populations including neutrophils and platelets after myeloblative treatments.

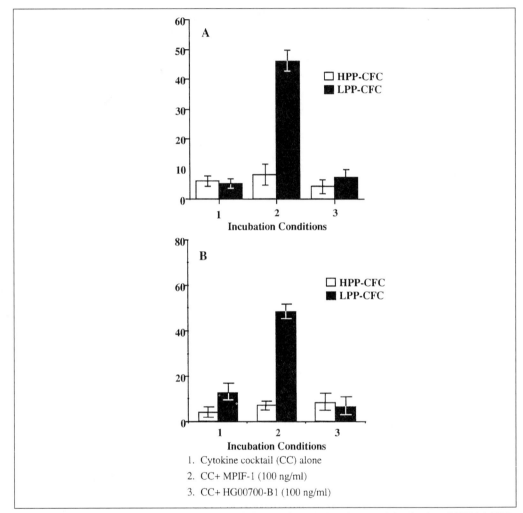

Fig. 4 MPIF-1 protects myeloid progenitors from 5-FU and Ara-c induced cytotoxicity *in vitro*. MPIF-1 inhibits growth of myeloid progenitors of both murine and human origin. To determine if the inhibitory effect of MPIF-1 can lead to the protection of LPP-CFC from the cytotoxicity of the cell cycle acting chemotherapeutic drug, lineage-depleted populations of cells (Lin⁻ cells) were isolated from mouse bone marrow and incubated in the presence of multiple cytokines with or without MPIF-1. After 48 hours, one set of each culture received 5-Fu and the incubation was then continued for an additional 24 hours, at which point the numbers of surviving LPP-CFC were determined by a clonogenic assay. As shown in the figure, about 40% of LPP-CFC were protected from the 5-Fu-induced cytotoxicity in the presence of MPIF-1, whereas little protection (< 5%) of LPP-CFC was observed in the absence of MPIF-1 or in the presence of an unrelated protein. HPP-CFC under the same culture conditions were not protected by MPIF-1, demonstrating the specificity of the MPIF-1 effect.

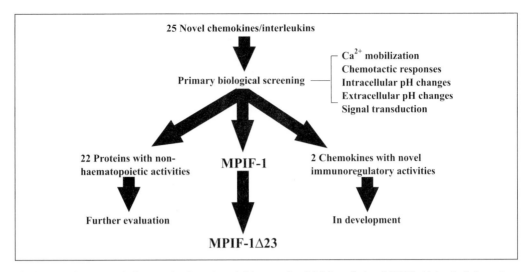

Fig. 5 Identification and characterization of myeloid progenitor inhibitory factor (MPIF-1). Using both homology searches and analysis of gene expression profiles on the EST database, a group of putative cytokine/interleukin genes were identified. Protein corresponding to each gene was expressed and analysed against a set of standardized functional screens. From the activity profiles obtained, clues towards further therapeutic potential of lead candidates were made. In subsequent and more specific assays, MPIF-1 and a recombinant derivative, MPIF-1Δ23, was shown to possess strong inhibitory activity against multipotential myeloid progenitor cells.

The genomics-based approach for identification and characterization of MPIF-1 was facilitated by discovery and testing of multiple novel chemokines and eliminating those with pleiotropic pro-inflammatory activities (Fig. 5).

The ability of MPIF-1 to function as a potent myeloprotectant without significant pro-inflammatory activity suggests that it may find therapeutic application as a chemoprotective agent and may spare early myeloid progenitors from the effects of commonly used chemotherapeutic drugs.

4. Expansion of the TNF/TNF receptor repertoire

The identity of a soluble factor responsible for cytotoxicity to tumours was first revealed in 1983 by B. B. Aggarwal and it was about ten years ago that two factors, TNF-α and LT (TNF-β), were isolated and the cDNA cloned (18, 19). Over the past few years, the list of novel genes belonging to the TNF ligand and TNF receptor superfamily has expanded rapidly (20–23).

Members of the TNF receptor (TNFR) family are Type I transmembrane glycoproteins and include a 55 kDa and 75 kDa form of the TNFR (24–26), CD40 (27), CD27 (28), CD30 (29), OX40 (30), a nerve growth factor receptor (31), 4-1BB (32), Fas (33), and several viral open reading frames that show significant identity with the extracellular domains of TNF family members (34). The receptors are all characterized by the presence of multiple cysteine-rich repeats of about 40 amino acids in the extracellular domain. In this region the homology between members is about

25%. The structures of the cytoplasmic domains are quite distinct. A short stretch of 80 amino acids, originally identified in the TNFRI and referred to as the 'death domain', has a significant homology to FAS/CD95 and several new members of this family.

Ligands for these receptors have been identified and belong to two cytokine superfamilies. The neurotrophins are basic nerve growth factor (NGF)-like dimeric molecules and include NGF, BDNF, NT-3, NT-4, and NT-5 (35, 36). The ligands of the TNF ligand superfamily are acidic, with approximately 20% sequence homology in the extracellular domains and exist mainly as membrane-bound Type II membrane proteins. Sequence alignments do show that there is a characteristic pattern of sequence conservation, with nine short regions of conserved sequences along the length of the molecule. The biologically active form is a trimeric/multimeric complex. Soluble forms of the TNF ligand superfamily have been identified for LT-α, which is always secreted, the TNF homotrimer, and FASL (37–40). Members of this family, until recently, have included TNF (37, 41–43), LT-α, LT-β, CD27L (44), CD30L (45), CD40L (46, 47), 4-1BB (48), OX40L (49), and FASL (38, 50). All members are involved in regulation of cell proliferation, activation, and differentiation including control of cell survival or death by apoptosis or cytotoxicity.

The advent of genomics-based gene discovery has contributed to the growing families of both novel members of the TNF/NGF receptor superfamily and the TNF ligand family. The receptor subfamily is characterized by multiple cysteine-rich domains that have been shown to be involved in ligand binding. This cysteine-rich

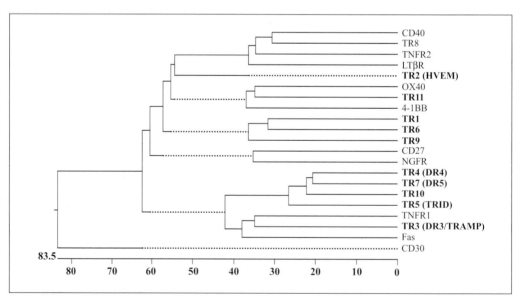

Fig. 6 Phylogenetic tree of TNF superfamily receptors. Alignment of TNF receptors by *Clustal* method using a PAM 250 matrix. The phylogenetic tree shows the ancestral relationship among the reported TNF receptors. Novel receptors identified by HGS are in bold. The length of each pair of branches represents the distance between sequence pairs, measured by number of non-conservative amino acid substitutions.

repeat provides one of the motifs that allow identification of novel receptors. This motif can be defined by the consensus pattern: C–x(4,6)–[FYH]–x(5,10)–C–x(2,3)–C–x(7,11)–C–x(4,6)–[DNEQSKP]–x(2)–C. In addition, the presence of a 'death domain' originally identified in the intracellular domain of TNFR, has provided another motif by which to search large databases. This domain is defined by a profile composed of the proteins known to contain this heterodimerization domain. The death domain is also found in a variety of other proteins including several involved in the death-signalling pathway transduced through members of the TNF family. Receptors recently identified by database analysis include TR1/AFPF/OPG (51), TR2/HVEM/ATAR (52, 53), TR3/DR3/WSL-1/APO3/TRAMP/LARD (54, 55), TR4/DR4 (56), TR5/TRID (57), TR6, TR7/DR5 (57), TR8 (Trance, Bank) (58, 59), TR9/DR6, TR10/TRUND (60), TR11. The phylogenetic similarities between these new receptors and the existing members of the TNFR superfamily are presented in Fig. 6.

Using the cysteine homologies present in the TNF ligand family, several novel ligands have been identified. The consensus for this family is [LV]–x–[LIVM]–x(3)–G–[LIVMF]–Y–[LIVMFY] (2)–x(2)–[QEKHL]–[LIVMGT]–x–[LIVMFY]. Ligands identified based on this motif include TL1/VEGI, TR2/TRAIL/APO2 (61), TL3/TNF-δ, TL4/TNF-ε, TL5/LIGHT, TL6/endokine, TL7/neutrokine. The phylogenetic similarities between the new TNF-like ligands and the existing members of the TNF ligand superfamily are presented in Fig. 7.

Through the use of genomic discovery, two new ligand receptor pairs have been identified. AIM II has been shown to bind the TR2/HVEM receptor and the LT-β receptor (62), joining TNF-α and LT-α in the property of binding multiple receptors. Also, TL2/TRAIL/APO2 has been shown to bind the DR4, DR5, and TRID receptors in addition to the newly described TRUND (56, 57, 60).

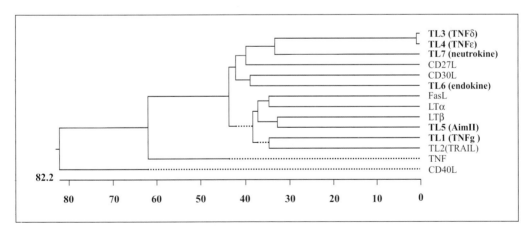

Fig. 7 Phylogenetic tree of TNF ligands. Alignment of TNF ligands by *Clustal* method using a PAM 250 matrix. The phylogenetic tree shows the ancestral relationship among the reported TNF ligands. Novel ligands identified by HGS are in bold. The length of each pair of branches represents the distance between sequence pairs, measured by number of non-conservative amino acid substitutions.

The discovery of the new family members representing receptors for TRAIL high-lights one of the major advantages of a database approach to gene discovery. While it might take an individual laboratory or several laboratories years to identify these novel ligand/receptor pairs, these genes were identified from the database in a relatively short period of time allowing scientists to experimentally determine ligand–receptor interactions and gain insight into the complex signal transduction mechanisms associated with these superfamily members. The quick expansion of novel TNFR members containing the 'death domain' represents a very compelling example. Until quite recently there were two TNFR family members known to contain the 'death domain' TNFR1 and FAS/APO (61). By expanding the EST database at HGS, several novel death domain containing receptors were identified (Fig. 8). Both DR4 and DR5 contain death domains in the cytoplasmic region as do TNFRI and FAS/APO. Unlike Fas, TNFR, and DR3, DR4 and DR5 do not use the death adapter molecule FADD (56, 57) to transmit the death signal. Activation of downstream caspases by DR5 may involve FLICE2 (57) whereas DR4 involves FLICE1 (56). TRID, however, is completely lacking a cytoplasmic tail and is attached to the membrane by a glycosyl phosphatidylinositol linkage (57). Therefore, TRID may serve as an antagonist decoy receptor. Expression of TRID actually protects cells from TRAIL-induced apoptosis, consistent with a role involved in cell protection. Recently, a new member of the TNFR family was identified that contains a cyto-

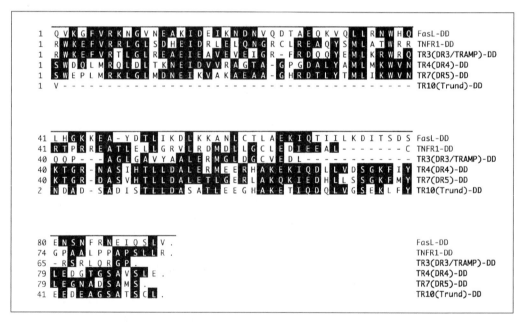

Fig. 8 Alignment of death domain region for TNF superfamily receptors. Alignment of death domain from cytoplasmic region of TNF receptors involved in apoptotic cell death signalling. Those receptors identified at HGS are in bold. Other common names for these receptors as they may now appear in the literature are in parenthesis.

plasmic tail but an incomplete death domain. This receptor, called TRUND (60), is incapable of transmitting a death signal, and may also act as a decoy receptor.

This example clearly demonstrates the power of a genetic approach to providing insight into a complex physiological response where TRAIL signalling *in vivo* is controlled by a family of receptors and a complex receptor mechanism exists to balance the positive and negative modulation of the death-signalling pathway and maintain homeostasis within the cell.

5. Database approaches toward high-throughput discovery of novel cytokines

Because our current knowledge of cytokines has been derived from a data set that probably corresponds to less than 10% of expressed human genes, it is clear that many more have yet to be discovered. Since the vast majority of expressed human genes have now been identified and deposited in EST databases, it would appear that the cytokine and secreted proteins that do exist bear little resemblance at the amino acid level to the cytokines identified to date. In the absence of homology, other methodologies will need to be developed if we are to use these large sequence data collections as a source for cytokine discovery. Approximately 30% of the database is represented by genes with no significant homology. One characteristic cytokines, growth factors, and other molecules that are directed through the cellular membrane have in common is a signal peptide at the 5′ terminus of the open reading frame. The signal sequence consists of three defined regions:

(a) *n*-regions that are often positively charged and adjacent to the initiator methionine and typically is 1–5 amino acids.

(b) An *h*-region of 7–15 amino acids represents a central hydrophobic stretch.

(c) A *c*-region of 3–7 amino acids that provides the cleavage site for a signal peptidase. This cleavage site is defined by a (–3, –1) (63) rule, where any small, uncharged residues are allowed.

While no explicit homology exists between various signal peptides, there does exist a bias for certain preferred residues in the three regions of the signal peptide. This fact can be utilized in a bioinformatics-based approach for selection of potential secreted candidates based on the properties of this motif. There are several programs available that can help to distinguish potential signal sequences, including programs based on multiple alignments (profile) (64), on patterns in the primary sequence generated by alignments of a family (Hidden Markoff Model) (65), and on the set of rules developed by Gunter von Heinje, PSORT (66) and *SignalP* (9).

Certainly for such a strategy to be successful, it must be established that the methodology is robust and discriminatory against non-secreted candidates. While these algorithms have proven to work quite well on data sets consisting of full-length, well annotated gene sequences, such as the Swiss protein database, the ability to be selective on an EST database needs to be addressed. This is easily accomplished

Table 1 Assessment of signal peptide selection criteria

Data set from Swiss Prot	No. of full-length candidates in HGS database	No. selected by *SignalP*
Proteins containing signal sequence	298	284 (95%)
Cytoplasmic proteins	077	008 (10%)
Nuclear proteins	096	002 (2%)
Type II membrane proteins	017	010 (59%)

by creating a test data set comprised of ESTs representing the 5′ ends of known genes with various cellular localizations. The results of such a test using *SignalP* are presented in Table 1. It can be seen that the algorithm is quite selective for secreted proteins and against both cytoplasmic and nuclear proteins, but has difficulty discriminating against Type II membrane proteins. This is not surprising, considering that the transmembrane region of Type II proteins possess all the characteristics of a signal peptide of an extracellular protein except the peptidase cleavage site. In addition, for a database approach to be comprehensive, the size of the data set must be large enough and the quality of the cDNA libraries used to create the data set of a high enough quality, to assure a high percentage of full-length cDNAs. Using a random sample of secreted proteins, the HGS database has cDNAs representing the 5′ end of over 75% of the test set. By extrapolation, one must assume that the database must also contain the 5′ end of a similar percentage of unknown secreted proteins.

Once a set of putative secreted factors is established, methodologies must be developed to proceed from cDNA to prediction of biological function. The most direct approach towards this end is to express the proteins coded by the cDNAs and test these proteins in a high throughput fashion on biologically relevant systems. This approach involves development of:

(a) Efficient methods of cloning large numbers of cDNAs in an expression system.

(b) Production of protein from this expression system.

(c) Biological testing of the proteins.

Independent of the expression system used, DNA must be prepared from the cloned cDNAs. There are many 96-well format plasmid purification systems available, including two robotic systems from Autogen and from Qiagen.

Production of protein can be accomplished using insect, yeast, or mammalian systems, where there will be a high likelihood that the protein will be expressed in the proper conformation. Transfection is completely amenable to a multiwell high-throughput plate format. Depending on the number of biological tests, the transfection can be scaled to produce sufficient quantities of conditioned supernatants. The cell type, cell number, transfection protocol, and DNA quantities must all be optimized. Various cell types are appropriate, dependent on optimization for maximum expressions. Several cell lines, including HEK293, COS7, CHO, or BHK, are

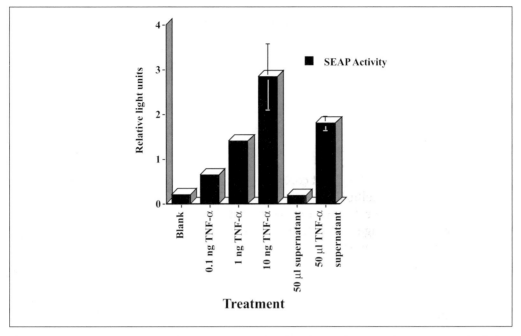

Fig. 9 Response of Jurkat: NFκB–SEAP cell line. Jurkat T cells stably expressing the secreted alkaline phosphatase (SEAP) reporter gene under the control of a synthetic NFκB promoter element were seeded in a 96-well dish (100 000 cells per well). The cells were then treated with either recombinant TNF-α or culture supernatant generated from 293T cells either mock transfected or transfected with a TNF-α expression vector. At 24 hours post-treatment, SEAP activity was determined using a 96-well Dynex plate luminometer.

useful. For transfection methods, any of the many cationic lipids work efficiently, but must be selected based on cell type and conditions must be optimized.

Biological testing of the supernatant can be based on a number of readouts and the strategy is dependent on the biological response or physiological state that one is interested in modulating. Systems based on heterologous reporters can be used to detect modulation of the expression levels of relevant indicator genes. For example, if the goal is the detection of novel factors that modulate immune response, a strategy based on the NFκB element directing transcription of a reporter system, such as secreted alkaline phosphatase (SEAP), can be built. This system would detect the activation of NFκB and subsequent binding to the enhancer element in the reporter construct as the dose-dependent expression of SEAP. Many other reporter genes are efficient including luciferase, β-galactosidase, and CAT; however, these reporters involve processing of cells to produce extracts. However, SEAP activity can be measured directly from the supernatant of the transfected cells (Fig. 9).

Other systems involving activation of surface markers can be based on FACS or ELISA assays. Instrumentation is now available that can detect changes in intracellular calcium ions, or pH, or membrane potential based on changes in the fluorometric emission spectrum of particular dyes. As an example, interaction of chemokines with the appropriate seven transmembrane G protein-coupled receptor elicits an increase in the calcium flux within the cell. In addition, secretion of particular cytokines or growth factors in response to particular signals can be used as a readout of biological activity. For example, detection of IFN-γ or IL-4 secretions by T cells can be used as a indicator of switching between Th1 and Th2 subsets of T helper cells.

References

1. Adams, M. D., Kelley, J. M., Gocayne, J. D., *et al.* (1991) Complementary DNA sequencing: expressed sequence tags and the human genome project. *Science*, **252**, 1651.
2. Matsubara, K. and Okubo, K. (1993) Identification of new genes by systematic analysis of cDNAs and database construction. *Curr. Opin. Biotechnol.*, **4**, 672.
3. Boguski, M. S., Lowe, J. M. J., and Tolstoshev, C. M. (1993) dbEST-database for 'expressed sequence tags'. *Nature Genet.*, **4**, 332.
4. Liew, C. C. (1993) A human heart cDNA library: the development of an efficient and simple method for automated DNA sequencing. *Cardiology*, **25**, 891.
5. Taskeda, J., Yano, H., Eng, S., Zeng, Y., and Bell, G. (1993) A molecular inventory of human pancreatic islets: sequence analysis of 1000 cDNA clones. *Hum. Mol. Genet.*, **2**, 1793.
6. Geiser, L. and Swaroop, A. (1992) Expressed sequence tags and chromosomal localization of cDNA clones from a subtracted retinal pigment epithelial library. *Genomics*, **13**, 873.
7. Khan, A., Wilcox, A., Polymeropolous, M., Hopkins, J., Stevens, T., Robinson, M., *et al.* (1992) *Genetics*, **2**, 180.
8. Sutton, G. G., White, O., Adams, M. D., and Kerlavage, A. R. (1995) TIGR Assembler: a new tool for assembling large shotgun sequencing projects. *Genome Sci. Technol.*, **1**(1), 9.
9. Nielsen, H., Engelbrecht, J., Brunak, S., and von Heijne, G. (1997) Identification of prokaryotic and eukaryotic signal peptides and prediction of their cleavage sites. *Protein Eng.*, **10**, 1.
10. Altschul, S. F., Gish, W., Miller, W., Myers, E. W., and Lipman, D. J. (1990) Basic local alignment search tool. *J. Mol. Biol.*, **215**, 403.
11. Rollin, B. J. (1997) Chemokines. *Blood*, **90**, 909.
12. Baggiolini, M., Dewald, B., and Moser, B. (1997) Human chemokines: an update. *Annu. Rev. Immunol.*, **15**, 675.
13. Wells, T. N. and Peitsch, M. C. (1997) The chemokine information source: identification and characterization of novel chemokines using the World Wide Web and expressed sequence tag databases. *J. Leuk. Biol.*, **61**, 545.
14. Patel, V. P., Kreider, B. L., Li, Y., *et al.* (1997) Molecular and functional characterization of two novel C–C chemokines as inhibitors of two distinct classes of myeloid progenitors. *J. Exp. Med.*, **185**, 1163.
15. Forssman, U., Delgado, M. B., Uguccioni, M., Loetscher, P., Garotta, G., and Baggiolini, M. (1997) Ckβ-8, a novel CC chemokine that predominantly acts on monocytes. *FEBS Lett.*, **408**(2), 211.
16. Patel, V. P., Kreider, B. L., Li, Y. *et al.* (1997). Molecular and functional characterization of two novel human C-C chemokines as inhibitors of two distinct classes of myeloid progenitors. *J. Exp. Med.*, **185**, 1163.
17. Manuscript in preparation.
18. Pennica, D., Nedwin, G. E., Hayflick, J. F., Seeburg, P. H., Derynen, R., *et al.* (1984) Human

tumor necrosis factor: precursor structure, expression and homology to lymphotoxin. *Nature*, **312**, 724.

19. Gray, P. W., Aggarwal, B. B., Benton, C. V., Bringman, T. S., Henzel, W. J., *et al.* (1984) Cloning and expression of cDNA for human lymphotoxin, a lymphokine with tumor necrosis activity. *Nature*, **312**, 721.

20. Aggarwal, B. B. and Natarajan, K. (1996) Tumor necrosis factors: developments during the last decade. *Eur. Cytokine Netw.*, **7**, 93.

21. Beutler, B. and van Huffel, C. (1994) Unraveling function in the TNF ligand and receptor families. *Science*, **264**, 657.

22. Armitage, R. J. (1994) Tumor necrosis factor receptor superfamily members and their ligands. *Curr. Opin. Immunol.*, **6**, 407.

23. Gruss, H.-J. and Dower, S. K. (1995) Tumor necrosis factor ligand superfamily: involvement in the pathology of malignant lymphomas. *Blood*, **85**(12), 3378.

24. Loetscher, H., Pan, Y.-C. E., Lahm, H.-W., *et al.* (1990) Molecular cloning and expression of the human 55 kDa tumor necrosis factor receptor. *Cell*, **61**, 351.

25. Schall, T. J., Lewis, M., Koller, K. J., *et al.* (1990) Molecular cloning and expression of a receptor for human tumor necrosis factor. *Cell*, **61**, 361.

26. Smith, C. A., Davis, T., Anderson, D., *et al.* (1990) A receptor for rumor necrosis factor defines an unusual family of cellular and viral proteins. *Science*, **248**, 1019.

27. Stamenkovic, I., Clark, E. A., and Seed, B. (1989) A B-lymphocyte activation molecule related to the nerve growth factor receptor and induced by cytokines in carcinomas. *EMBO J.*, **8**, 1403.

28. Camerini, D., Walz, G., Loenen, W. A. M., Borst, J., and Seed, B. (1991) The T cell activation antigen CD27 is a member of the NGF/TNF receptor gene family. *J. Immunol.*, **147**, 3165.

29. Dürkop, H., Latza, U., Hummel, M., Eitelbach, F., Seed, B., and Stein, H. (1992) Molecular cloning and expression of a new member of the nerve growth factor receptor family that is characteristic for Hodgkin's disease. *Cell*, **68**, 421.

30. Mallet, S., Fossum, S., and Barclay, A. N. (1990) Characterization of the MRC OX40 antigen of activated CD4 positive T lymphocytes—a molecule related to nerve growth factor receptor. *EMBO J.*, **9**, 1063.

31. Johnson, D., Lanahan, A., Buck, C. R., *et al.* (1986) Expression and structure of the human NGF receptor. *Cell*, **47**, 545.

32. Kwon, B. S. and Weissman, S. M. (1989) cDNA sequences of two inducible T-cell genes. *Proc. Natl. Acad. Sci. USA*, **86**, 1963.

33. Itoh, N., Yonehara, S., Ishii, A., *et al.* (1991) The polypeptide encoded by the cDNA for human cell surface antigen fas can mediate apoptosis. *Cell*, **66**, 233.

34. Banchereau, J., Bazan, F., Blanchard, D., *et al.* (1994) The CD40 antigen and its ligand. *Annu. Rev. Immunol.*, **12**, 881.

35. Bradshaw, R. A., Blundell, T. L., Lapatto, R., McDonald, N. Q., and Murray-Rust, J. (1993) Nerve growth factor revisited. *Trends Biochem. Sci.*, **18**, 48.

36. Eide, F. F., Lowenstein, D. H., and Reichardt, L. F. (1993) Neurotrophins and their receptors—current concepts and implications for neurologic disease. *Exp. Neurol.*, **121**, 200.

37. Pennica, D., Nedwin, G. E., Hayflick, J. S., Seeburg, P. H., Derynck, R., *et al.* (1984) Human tumour necrosis factor: precursor structure, expression and homology to lymphotoxin. *Nature*, **312**, 724.

38. Suda, T., Takahashi, T., Golstein, P., and Nagata, S. (1993) Molecular cloning and

expression of the Fas ligand, a novel member of the tumor necrosis factor family. *Cell*, **75**, 1169.

39. Kriegler, M., Perez, C., DeFay, K., Albert, I., and Lu, S. D. (1988) A novel form of TNF/cachectin is a cell surface cytotoxic transmembrane protein: ramifications for the complex physiology of TNF. *Cell*, **53**, 45.

40. Mohler, K. M., Sleath, P. R., Fitzner, J. N., Cerretti, D. P., Alderson, M., *et al.* (1994) Protection against a lethal dose of endotoxin by an inhibitor of tumour necrosis factor processing. *Nature*, **370**, 218.

41. Old, L. J. (1985) Tumor necrosis factor (TNF). *Science*, **230**, 630.

42. Wang, A. M., Creasey, A. A., Ladner, M. B., Lin, L. S., Strickler, J., *et al.* (1985) Molecular cloning of the complementary DNA for human tumor necrosis factor. *Science*, **228**, 149.

43. Beutler, B. and Cerami, A. (1986) Cachectin and tumour necrosis factor as two sides of the same biological coin. *Nature*, **320**, 584.

44. Goodwin, R. G., Alderson, M. R., Smith, C. A., *et al.* (1993) Molecular and biological characterization of a ligand for CD 27 defines a new family of cytokines with homology to tumor necrosis factor. *Cell*, **73**, 447.

45. Smith, C. A., Gruss, H.-J., Davis, T., *et al.* (1993) CD30 antigen, a marker for Hodgkin's lymphoma, is a receptor whose ligand defines an emerging family of cytokines with homology to TNF. *Cell*, **73**, 1349.

46. Armitage, R. J., Fanslow, W. C., Strockbine, L., *et al.* (1992) Molecular and biological characterization of a murine ligand for CD40. *Nature*, **357**, 80.

47. Graf, D., Korthauer, U., Mages, H. W., Senger, G., and Kroczek, R. A. (1992) Cloning of TRAP, a ligand for CD40 on human T cells. *Eur. J. Immunol.*, **22**, 3191.

48. Goodwin, R. G., Din, W. S., Davis-Smith, T., *et al.* (1993) Molecular cloning of a ligand for the inducible T-cell gene 4–1BB: a member of an emerging family of cytokines with homology to tumor necrosis factor. *Eur. J. Immunol.*, **23**, 2631.

49. Baum, P. R., Gayle, R. B. III, Ramsdell, F., *et al.* (1994) Molecular characterization of murine and human OX40/OX40 ligand systems: identification of a human OX40 ligand as the HTLV-1 regulated protein gp34. *EMBO J.*, **13**, 3992.

50. Lynch, D. H., Watson, M. L., Alderson, M. R., *et al.* (1994) The mouse Fas-ligand gene is mutated in *gld* mice and is part of a TNF family gene cluster. *Immunity*, **1**, 131.

51. Simonet, W. S., *et al.* (1997) Osteoprotegerin: a novel secreted protein involved in the regulation of bone density. *Cell*, **89**(2), 309.

52. Montgomery, R. I., Warner, M. S., Lum, B. J., and Spear, P. G. (1996) Herpes simplex virus-1 entry into cells mediated by a novel member of the TNF/NGF receptor family. *Cell*, **87**(3), 427.

53. Kwon, B. S., Tan, K. B., Ni, J., *et al.* (1997) A newly identified member of the tumor necrosis factor receptor superfamily with a wide tissue distribution and involvement in lymphocyte activation. *J. Biol. Chem.*, **272**(22), 14272.

54. Chinnaiyan, A. M., O'Rourke, K., Yu, G. L., *et al.* (1996) Signal transduction by DR3, a death domain-containing receptor related to TNFR-1 and CD95. *Science*, **174**(5289), 990.

55. Kitten, J., Raven, T., Jiang, Y. P., *et al.* (1996) A death-domain-containing receptor that mediates apoptosis. *Nature*, **384**(6607), 372.

56. Pan, G., O'Rourke, K., Chinnaiyan, A. M., *et al.* (1997) The receptor for the cytotoxic ligand TRAIL. *Science*, **276**, 111.

57. Pan, G., Ni, J., Wei, Y.-W., *et al.* (1997) An antagonist decoy receptor and a death domain-containing receptor for TRAIL. *Science*, **277**, 815.

58. Wong, B. R., Rho, J., Arron, J., *et al.* (1997) TRANCE is a novel ligand of the tumor necrosis

factor receptor family that activates c-Jun N-terminal kinase in T cells. *J. Biol. Chem.*, **272**(40), 25190.

59. Anderson, D. M., Maraskovsky, E., Billingsley, W. L., *et al.* (1997) A homologue of the TNF receptor and its ligand enhance T cell growth and dendritic-cell function. *Nature*, **390**, 175.

60. Pan, G., Ni, J., Yu, G.-L., Wei, Y.-F., and Dixit, V. M. (1998) TRUNDD, a new member of the TRAIL receptor family that antagonizes TRAIL signalling. Submitted.

61. Wiley, S. R., Schooley, K., Smolak, P. J., *et al.* (1995) Identification and characterization of a new member of the TNF family that induces apoptosis. *Immunity*, **6**, 673.

62. Mauri, D. N., Ebner, R., Kochel, K. D., *et al.* (1998) LIGHT, a new member of the TNF superfamily and lymphotoxin (LT)a are ligands for herpesvirus entry mediator (HVEM). *J. Immunol.*, in press.

63. von Heijne, G. (1986) A new method for predicting signal sequence cleavage sites. *Nucleic Acids Res.*, **14**(11), 4683.

64. Grisbskov, M. (1987) Profile analysis: detection of distantly related proteins. *Proc. Natl. Acad. Sci. USA*, **84**(13), 4355.

65. Eddy, S. R. (1995) Multiple alignment using hidden Markov models. In *Proceedings of the Third International Conference on Intelligent Systems for Molecular Biology* (ed. C. Rawlings, *et al.*), p. 114. AAAI Press, Menlo Park.

66. Nakai, K. and Kanehisa, M. (1992) A knowledge base for predicting protein localization sites in eukaryotic cells. *Genomics*, **14**, 897.

2 | Cytokine signal transduction

KEATS NELMS

1. Introduction

Cytokines are important mediators of cellular communication. The synthesis and release of a cytokine by one cell represents a signal that can elicit dramatic responses in cells that express the corresponding cytokine receptors. Two major challenges arise in trying to understand this type of cellular communication. The first is to fully understand the processes that initiate and result in cytokine production. The second challenge is to understand how individual cytokines evoke different cellular responses upon receptor engagement. This chapter examines the latter problem by focusing on the intracellular signalling pathways that lead to different cellular responses.

2. Initiation of signal transduction by cytokine receptors

Two key events are required to initiate the intracellular signalling pathways activated by cytokines. The first is the binding of cytokine by receptor molecules that mediate the transduction of signals from the extracellular environment into the cytoplasm. The second is the activation of tyrosine kinases that catalyse the phosphorylation of receptor and signalling molecules. Such phosphorylation reactions facilitate the interaction of receptors and signalling molecules and represent a form of molecular communication through which intracellular signals are delivered.

2.1 Cytokine receptors

Cytokine receptors are composed of two functionally distinct domains, an extracellular cytokine-binding domain and a cytoplasmic domain that is the initiation site of intracellular signalling processes. Based on conserved structural characteristics in the extracellular domain, cytokine receptors can be divided into four subclasses (reviewed in refs 1 and 2). Type I receptors include the receptors for a number of interleukins (IL) as well as granulocyte-macrophage colony-stimulating factor (GM-CSF), erythropoietin, growth hormone, and prolactin. Type II cytokine receptors include the interferon and IL-10 receptors. Type III receptors are represented by the tumour necrosis factor receptors whereas Type IV receptors are represented by IL-1 receptors.

Regardless of subtype, cytokine receptors share several common characteristics. First, cytokine receptors usually function as multisubunit complexes. Receptors for different cytokines may even share common subunits, such as the IL-2, IL-4, IL-7, IL-9, and IL-15 receptors that each form a functional complex with a common subunit termed the gamma common chain (γ_c) (2). Secondly, the cytoplasmic domains of different cytokine receptors may contain subdomains that are critical to the activation of distinct cellular responses, such as proliferation or differentiation. Examples of such subdomains have been delineated in the IL-2 and IL-4 receptors (2, 3). These subdomains contain specific tyrosine residues that are phosphorylated upon receptor engagement and thus become sites of interaction with signalling molecules. Thirdly, cytokine receptor cytoplasmic domains have no endogenous tyrosine kinase activity capable of catalysing receptor phosphorylation. In fact, the tyrosine kinase activity induced by cytokine binding comes from kinases associated with receptor cytoplasmic domains.

2.2 Activation of receptor-associated tyrosine kinases

Cytokines initiate intracellular signals through the binding and oligomerization of receptor subunits. Cytokine binding can result in the heterodimerization of receptor subunits (as with receptors that share the γ_c chain), receptor homodimerization (as with the growth hormone receptor), or receptor trimerization (as with the tumour necrosis factor receptor). Yet the common outcome of each receptor oligomerization event is the activation of tyrosine kinases associated with receptor cytoplasmic domains followed by phosphorylation of cellular substrates (Fig. 1). Two families of tyrosine kinases in particular have been implicated in cytokine signal transduction — the JAK and Src kinases.

The JAK kinases (named after Janus, the Roman god with two faces) have two catalytic domains and have been implicated in the signalling of both Type I and Type II cytokine receptors (Fig. 2A). To date, four JAK kinases have been described including JAK1, JAK2, JAK3, and Tyk2 (reviewed in ref. 4). JAK kinases associate with the juxtamembrane region of receptor subunit cytoplasmic tails and are thought to become activated by cross-phosphorylation after cytokine-induced oligomerization of receptor subunits (Fig. 1A). Although different JAK kinases appear to associate specifically with different receptors, it is not clear if specific JAK kinases are required for the initiation of particular intracellular signals. It is clear, however, that JAK kinases in general are critical for the activation of a specific signalling pathway involving signal transducers and activators of transcription (STAT) (discussed in Section 3.3.1).

In contrast to JAK kinases that function specifically in cytokine receptor signalling, Src kinases have been implicated in numerous cellular processes. However, several members of the Src family, including Src, Lck, Fyn, and Lyn, also have been

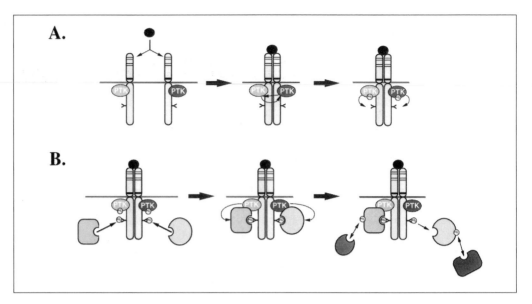

Fig. 1 A generalized model for cytokine signal transduction. (A) Cytokine signalling is initiated by binding of the cytokine to its receptor (a Type I cytokine receptor is shown here) resulting in cross-linkage of receptor subunits, activation of receptor-associated kinases, and receptor tyrosine phosphorylation. (B) Cytokine receptor phosphorylation results in the association or receptors with different signalling molecules.

implicated in cytokine signalling. Like JAK kinases, Src-family kinases have a highly conserved structure including a C-terminal kinase domain and N-terminal domains, termed the Src-homology domains 2 (SH2) and 3 (SH3), that are critical for the interaction of these kinases with other molecules (Fig. 2A). Many substrates of cytokine-activated Src kinases have yet to be defined; however, Src kinases such as Lck are important in the phosphorylation of specific receptor subunits (2) (Fig. 1A). Indeed, it is this type of receptor phosphorylation event that results in the inter-action of receptors with signalling molecules that specifically recognize and bind phosphotyrosine residues. Once bound to phosphorylated receptors, these signalling molecules themselves become phosphorylated by receptor-associated kinases, enabling them to interact with additional signalling molecules (Fig. 1B). Thus, receptor oligomerization and kinase activation results in a series of molecular interactions that form an intracellular signalling pathway. In the end, the cellular response activated by each pathway is determined by the individual functions of its molecular components.

3. Cytokine signalling pathways

Although the biological effects elicited by different cytokines can vary dramatically, it is remarkable that the individual signalling pathways activated are often in-distinguishable. Ultimately, the unique response elicited by a cytokine is a reflection

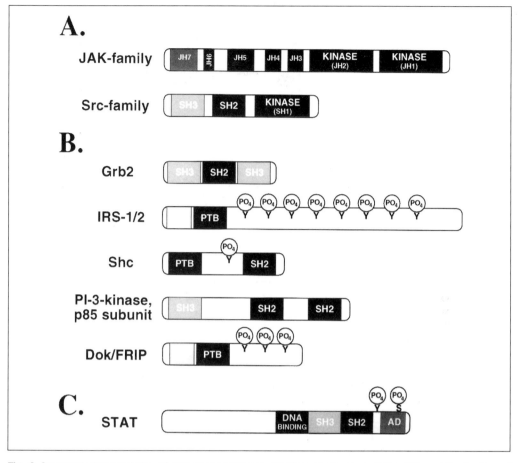

Fig. 2 Structural characteristics of signalling molecules. (A) General structure of the JAK- and Src-families of cytokine receptor-associated kinases. Domains homologous between different JAK (JH) or Src (SH) kinases are numbered and indicated. (B) Adapter molecules involved in cytokine signalling pathways. SH2, SH3, PH, and PTB domains are involved in protein to protein interactions and are defined in the text. Sites of tyrosine (Y) phosphorylation (PO_4) are indicated. (C) General structure of STAT molecules. Protein domains involved in DNA binding and transcriptional activation (AD) are indicated. Sites of tyrosine (Y) or serine (S) phosphorylation (PO_4) are indicated.

of the array of signalling pathways that it activates and of the interplay between these pathways. It is clear that cellular responses induced by cytokine stimulation may be categorized into several general types. These include cellular proliferation, activation of specific genes and, in some cases, cell death. Signalling pathways that result in cellular responses to cytokine stimulation have been identified and emerging evidence strongly suggests that these pathways are not mutually exclusive but intersect at a growing number of control points. The following section defines four key signalling pathways that are activated in response to cytokine receptor engagement.

3.1 The Ras/mitogen activated protein kinase pathway

One of the best studied and most complex signalling pathways is the Ras/mitogen activated protein kinase (MAPK) pathway. This pathway is highly conserved between organisms as varied as fruit flies, worms, and mammals. Although cellular proliferation or differentiation may occur upon activation of the Ras/MAPK pathway (depending on the cellular context and activating stimuli), cytokine activation of this pathway is most often associated with cellular proliferation and with the prevention of apoptosis. Many cytokine receptors, including all of the haematopoietin receptor family, have been demonstrated to activate the Ras/MAPK pathway upon cytokine binding (5, 6). Interestingly, certain cytokines may only activate this pathway in particular cell types. For example, IL-4 activates the Ras/MAPK pathway in B cells and keratinocytes but fails to activate it in a number of haematopoietic cell types (7–9). Thus, the specificity of Ras/MAPK pathway activation may be related to the cell type as well as the activating cytokine.

The Ras/MAPK pathway was one of the first pathways in which extracellular signals were biochemically and genetically linked to activation of gene expression in the nucleus (10) (Fig. 3). The central components of this pathway are a series of related small (21 kDa) GTPases, the best characterized of which are the p21 Ras proteins encoded by the Ha-, Ki-, and N-*ras* oncogenes. Additional small GTPases that are related to Ras, including the Rho, Rac, and Rap1 proteins, are involved in distinct pathways that share many of the characteristics of the Ras/MAPK pathway (11–13). Rho and Rac, in particular, may play a role in inflammatory cytokine signalling (11). However, the relative importance of these Ras-related proteins in cytokine signalling has yet to be fully determined.

Ras proteins undergo post-translational modification by isoprenylation, which localizes the Ras proteins to the plasma membrane (14). Membrane-associated Ras binds GTP, which is then hydrolysed to GDP by the GTPase activity of Ras. In unstimulated cells, the dissociation rate of GDP from Ras is very slow causing most cellular Ras to be in the Ras–GDP form. Cytokine stimulation results in a rapid increase in the amount of Ras–GTP, the biologically active form that initiates a downstream cascade of kinase reactions termed the MAPK pathway (15). Thus, control of activated Ras–GTP accumulation is critical to the regulation of the Ras/MAPK pathway.

3.1.1 Regulation of Ras activation

Two classes of proteins control the cellular ratio of Ras–GTP to Ras–GDP and thus regulate the activation of the MAPK pathway. One class, the Ras GTPase activating proteins, increase Ras GTPase activity several orders of magnitude and thus lead to the accumulation of inactive Ras–GDP (13, 16). The best characterized protein in this class is a 120 kDa protein referred to as RasGAP (for Ras GTPase activating protein). RasGAP contains two Src-homology 2 (SH2) domains, which bind protein sequences containing phosphotyrosine residues, as well as a central Src-homology 3 (SH3) domain that interacts with proline-rich sequences (13, 17). A second Ras GTPase activating protein, termed NF1 or neurofibromin, has a high degree of homology to the RasGAP catalytic domain but lacks SH2 or SH3 domains. Mutations in the *NF1* gene have been implicated in the inherited disease neurofibromatosis, which is characterized by a high incidence of tumours of neural crest origin (13). Although the role of NF1 in normal Ras regulation is not clear, mice lacking the *NF1* gene are embryonic lethal, as are knock-out mice lacking the RasGAP gene (18, 19).

The second class of proteins involved in regulating Ras activation are the guanine nucleotide exchange proteins that catalyse the exchange of GDP bound by Ras for GTP, which results in the formation of active Ras–GTP. Members of this class of proteins were first characterized utilizing yeast and *Drosophila melanogaster* genetic systems. The best characterized member of this class of proteins, termed Sos (Son Of Sevenless), was identified because of its involvement in the development of *D. melanogaster* photoreceptor cells, a process that also requires the tyrosine kinase *Sevenless* (20). Two mammalian homologues, Sos1 and Sos2, have been identified based on sequence similarity to the *D. melanogaster* Sos protein (21). Although the precise mechanism by which Sos functions has not been defined, the interaction of Sos with Ras is considered to be the primary mechanism by which Ras–GTP is formed in mammalian cells. Sos is constitutively associated with a second signalling molecule termed Grb2 (growth factor receptor-binding protein 2) and, as will be discussed, this association is critical for the ability of Sos to regulate Ras activation (22). Additional guanine nucleotide exchange proteins have been identified including CDC25 and C3G. In contrast to Sos, C3G does not activate Ras but does activate the related protein, Rap1 (12). Cell type-specific guanine nucleotide exchange proteins also exist. In particular, the T cell-specific protein Vav has been demonstrated to act as a guanine nucleotide exchange factor (23). The inherent activity of Sos does not appear to be altered by cytokine stimulation (20). Indeed, the critical event leading to the accumulation of Ras–GTP appears to be the accumulation of Sos near Ras at the plasma membrane, which is accomplished through the action of adapter proteins.

3.1.2 Adapter proteins in the Ras/MAPK pathway

The general function of adapter proteins is to act as a bridge between activated receptor molecules and proteins with some intrinsic catalytic activity. These adapter proteins do not have enzymatic activity but they do contain a number of recognizable domains found in different signalling molecules. In the case of the Ras/MAPK pathway, a group of well-characterized adapter proteins link Sos to activated receptor complexes and increase the amount of Sos at the plasma membrane where Ras is localized, thus facilitating the accumulation of biologically active Ras–GTP. While the general function of individual adapter proteins are similar, they also have unique characteristics that impact on their overall function and result in the regulation of signalling pathways other than the Ras/MAPK pathway. Thus, they will be discussed individually.

3.1.3 Grb2

Because of the constitutive association of Grb2 with Sos in mammalian cells, this is an adapter protein that plays a critical role in linking Sos to activated receptor molecules. The Grb2 protein consists of N-terminal and C-terminal SH3 domains that mediate the interaction with proline-rich sequences in Sos (17) (Fig. 2B). The SH3 domains flank an internal SH2 domain that is capable of binding to phosphotyrosine residues in receptors and other signalling molecules (17). A similar adapter, termed Crk, also has been shown to interact constitutively with the alternate guanine nucleotide exchange factor C3G (12). The importance of Grb2 to the Ras/MAPK pathway is underscored by genetic experiments that indicate that analogous proteins, Drk in *D. melanogaster* and SEM-5 in *Caenorhabditis elegans*, are required for Ras activation (24–26).

Grb2 has been shown to bind directly to activated growth factor receptors such as the EGF receptor, thus bringing Sos to the plasma membrane (17). However, in the activation of the Ras/MAPK pathway through cytokine receptors, Grb2/Sos rarely binds directly to the receptor molecules but is usually linked indirectly through an intermediate adapter molecule. Three adapter molecules linking the Grb2/Sos complex to cytokine receptors termed, Shc, IRS-1, and IRS-2, have been well characterized.

3.1.4 Shc

A defining characteristic of the adapter protein Shc is that it contains two distinct domains capable of binding tyrosine-phosphorylated receptor sequences. The C-terminal region of the Shc protein contains an SH2 domain that recognizes phosphotyrosine in the context of protein sequences C-terminal to the tyrosine while the N-terminal domain contains a phosphotyrosine binding (PTB) domain (17, 27) (Fig. 2B). The PTB domain, in contrast to SH2 domains, binds to phosphoprotein sequences based on the amino acid residues N-terminal to the phosphotyrosine residue (28). The PTB and SH2 domains of Shc mediate its interaction with phosphorylated receptor molecules. Once this interaction occurs, Shc itself is phosphorylated at Tyr317 and this phosphorylated tyrosine then serves as a docking site for the SH2 domain of Grb2 (29, 30). Thus, Shc links the Grb2/Sos complex to phosphorylated receptors and leads to Ras activation. Cytokine stimulation of haematopoietic cells, and IL-3 stimulation in particular, induces Shc phosphorylation at residues Tyr239 and Tyr240 (31). This phosphorylation is not required for Ras activation but does contribute to prevention of apoptosis by IL-3. Multiple cytokines have been demonstrated to activate Shc phosphorylation although the phosphorylation induced by IL-4 appears to be restricted to certain cell types (6, 8, 9). In cells in which IL-4 does not induce Shc phosphorylation, IL-4 has been demonstrated to stimulate the phosphorylation of two other related adapter proteins, IRS-1 and IRS-2.

3.1.5 Insulin receptor substrate 1 and 2

Insulin receptor substrate 1 (IRS-1) was first identified by White and colleagues as a highly phosphorylated substrate of the insulin and insulin-like growth factor receptors (32). The 185 kDa IRS-1 is related to a second molecule, IRS-2, that was first identified as the major substrate of the IL-4 receptor in haematopoietic cells (33–35). Both IRS-1 and IRS-2 are multiply phosphorylated in response to stimulation by a number of cytokines including IL-2, IL-4, IL-7, IL-9, and IL-15 (36) (Fig. 2B). IRS-1/IRS-2 molecules bind to phosphorylated receptors through a N-terminal PTB domain, which is identical in structure to the PTB domain of Shc (37, 38). The N-terminal region of IRS-1/IRS-2 also contains a pleckstrin homology (PH) domain similar to those found in a large number of signalling molecules. Although the function of the PH domain is still being defined, it may play a role in localizing proteins to the plasma membrane by interacting directly with phospholipids (39). IRS-1/IRS-2 molecules each have approximately 20 potential sites of tyrosine phosphorylation (32, 40). Some of these sites have been demonstrated to be bound by specific SH2 domains, which indicates that IRS-1 and IRS-2 act as cytosolic docking proteins capable of linking a variety of SH2 domain-signalling molecules to phosphorylated receptors (32, 33, 40). In particular, phosphorylated IRS-1/IRS-2 has been shown to interact with Grb2/Sos, linking them to membrane-bound Ras (33). Activation of the Ras/MAPK pathway is, however, only one aspect of IRS-1/IRS-2 function as a number of other pathways may be regulated through the interaction of different signalling molecules with phosphorylated IRS-1/IRS-2 (33).

3.1.6 Downstream effects of Ras activation: the MAPK cascade

Regulation of Ras–GTP accumulation is a central control point of the Ras/MAPK pathway. Once Ras–GTP has formed, it initiates the MAPK pathway that consists of a series of discrete kinase activation events and leads from the plasma membrane to the nucleus (Fig. 3). Ras-related proteins such as Rho and Rac appear to initiate kinase activation cascades analogous to the MAPK pathway, but distinct kinases are involved. It remains to be resolved to what degree the kinase cascades activated by different small GTPase proteins influence each other and whether different cytokines use different pathways preferentially. Also, the identity of all the cellular substrates of kinases in these pathways have not been delineated. However, it is clear that the activation of the MAPK pathway by Ras results in the phosphorylation of nuclear transcription factors. These nuclear phosphorylations are critical downstream events in the Ras/MAPK pathway.

A critical early event in the MAPK pathway is the activation of the serine/threonine kinase, Raf-1, resulting from the interaction of Raf-1 with activated Ras–GTP. Although the direct interaction of Raf-1 with Ras–GTP is required for the activation of Raf-1, this interaction alone does not activate Raf-1 (41, 42). Indeed, it is not completely understood how the interaction Raf-1 with Ras–GTP results in Raf-1 activation. It is possible that association with Ras–GTP may lead to Raf-1 tyrosine phosphorylation as evidenced by studies in which Ras oncogenicity is enhanced by the co-expression of tyrosine kinases that phosphorylate Raf-1 (43). Also, stimulation of T cells with IL-2 also results in the tyrosine phosphorylation of Raf-1, yet it is not clear if these phosphorylation events are critical to Raf-1 activation (44). Serine/threonine phosphorylation may play a more important role in Raf-1 activation

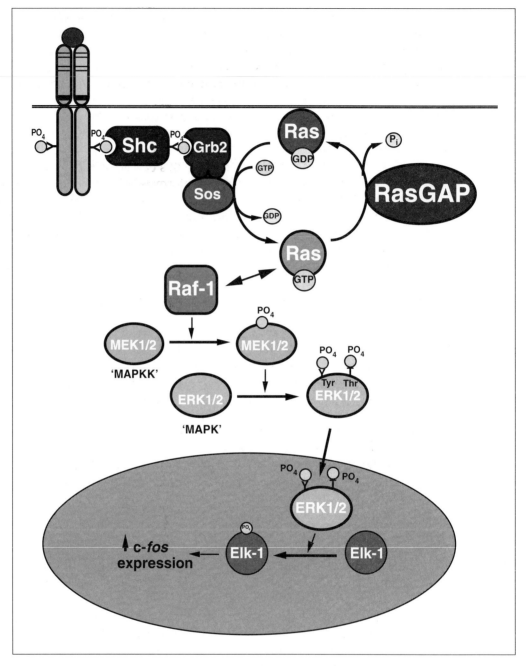

Fig. 3 The Ras/mitogen activated protein kinase (MAPK) pathway. The pathway is depicted as leading from a generalized cytokine receptor and includes the MAPK cascade that results in activation of c-*fos* gene expression. Tyrosine (Y) or threonine (T) phosphorylation (PO$_4$) are indicated.

because, upon stimulation, Raf-1 is phosphorylated on multiple serine/threonine residues. Indeed, phosphorylation by protein kinase C leads to Raf-1 activation (45). It has been suggested that activation of Raf-1 also may be influenced by the cytoplasmic protein 14-3-3 that interacts with Raf-1 in its inactive as well as its serine/threonine phosphorylated state (46). However, the relative importance of phosphorylation and 14-3-3 interactions to Raf-1 activation remains to be fully elucidated.

Activated Raf-1 phosphorylates and activates kinase intermediates of the Ras/MAPK pathway, referred to in general as MAPK kinases (MAPKK). These are dual specificity kinases with both serine/threonine and tyrosine kinase activities that phosphorylate and activate the key kinase in this cascade, referred to as MAPK. In the Ras/MAP kinase cascade, two specific isoforms of MAPKK have been identified and are termed MEK1 and MEK2 (for MAP/Erk kinases) (46). Activation of the MEKs results in the phosphorylation and activation of MAPK. As with MAPKK, two isoforms of MAPK have been identified and are referred to as ERK1 and ERK2 (for extracellular regulated kinases) (46). Once activated, these ERKs translocate to the nucleus where one of their primary activities is the phosphorylation of nuclear transcription factors. ERKs in particular catalyse the phosphorylation of the transcription factor Elk-1, resulting in the activation of the c-*fos* proto-oncogene (47) (Fig. 3).

Distinct kinases with functions similar to the ERKs can be activated through kinase cascades initiated by Ras as well as other small GTPases. These kinases include Jun nuclear kinases (JNK), which phosphorylate the transcription factor Jun, p38/CSBP kinase, which phosphorylates the transcription factor ATF2, and p90rsk, which phosphorylates Fos (46, 47). Signals leading to the activation of these different transcription factors are often co-ordinated. Indeed, the activation of the c-*fos* gene by ERK-activated Elk-1 and the activation of Jun by JNK phosphorylation leads to the formation of AP-1, a transcription activator composed of Fos and Jun heterodimers that is involved in the transcription of numerous genes (47). Ultimately, it is this induction of gene expression that leads to many of the cellular responses attributed to activation of the Ras/MAPK pathway.

3.2 The phosphoinositide-3-kinase pathway

Many aspects of the phosphoinositide-3-kinase (PI-3-kinase) pathway, a second signalling pathway activated by cytokines, have proven to differ strikingly from the Ras/MAPK pathway. Even so, the activation of PI-3-kinase is also clearly important in the stimulation of cellular proliferation by numerous cytokines including IL-2, IL-3, IL-4, IL-5, and GM-CSF and a number of the molecular events leading to the activation of PI-3-kinase by cytokines have now been elucidated (48–50).

Several biochemically distinct forms of PI-3-kinase exist, but the form of PI-3-kinase widely recognized to be activated by cytokines is a heterodimer composed of a 110 kDa catalytic (p110) and a 85 kDa regulatory (p85) subunit. The p85 regulatory subunit contains distinct domains that enable it to function as an adapter molecule linking the PI-3-kinase heterodimer to different signalling molecules (Fig. 2B). These domains include an N-terminal SH3 domain and two proline-rich regions that interact with SH3 domains in other signalling molecules (51). The C-terminal region of p85 contains tandem SH2 domains that link the p85/p110 heterodimer to tyrosine-phosphorylated proteins such as the adapter molecules IRS-1 and IRS-2 as well as activated receptor molecules (32, 40). The end-result of these interactions is to bring PI-3-kinase to the membrane near activated receptor complexes.

The tandem SH2 domains of the p85 subunit are separated by a 104 amino acid domain that mediates the interaction with the N-terminal region of the p110 catalytic subunit (52, 53). The interaction of p85 with p110 enhances the enzymatic activities of p110, which include phosphorylation of proteins on serine and threonine residues and the phosphorylation of phosphoinositide lipids (54). The functional significance of the serine/threonine kinase activity of the p110 catalytic subunit is not completely understood. However, phosphorylation of p85 by p110 results in an 80% reduction in PI-3-kinase activity, strongly suggesting that PI-3-kinase activity may be regulated through autophosphorylation (54). In addition to p85, other proteins have been shown to be targets of the serine/threonine kinase activity including the IRS-1 adapter molecule (55).

In contrast to the serine/threonine kinase activity, the lipid kinase activity of the p110 subunit has been more clearly defined and has been shown to mediate the transfer of phosphate from ATP to the D3 position of inositol in phosphatidylinositol, which results in one of three potential products: phosphatidylinositol (3)-phosphate [PtdIns(3)P], phosphatidylinositol (3,4)-bisphosphate [PtdIns(3,4)P_2], or phosphatidylinositol (3,4,5)-triphosphate [PtdIns(3,4,5)P_3] (56). In most cells PtdIns(3,4,5)P_3 production peaks within seconds of cellular stimulation followed by PtdIns(3,4)P_2. The rapid production of D3 phosphoinositide lipids catalysed by PI-3-kinase in response to mitogenic stimuli has led to the proposal that these lipids are biologically important second messenger molecules that contribute to the proliferative response (57).

Most cytokines that induce cellular proliferation have been demonstrated to activate PI-3-kinase in response to receptor engagement. Although the molecular details of this activation may differ between different cytokines, a common step in the activation of PI-3-kinase by cytokines involves the recruitment and interaction of p85 with components of the cytokine receptor signalling complex (Fig. 4). The most direct interaction of this type occurs between the SH2 domains of p85 and phosphorylated cytokine receptors such as the IL-2R β chain (58). More commonly, the p85 subunit is linked to cytokine receptors through indirect interactions with other proteins. Indeed, two additional interactions localize PI-3-kinase to cytokine receptor complexes. One of these involves the interaction of p85 subunit with receptor-associated adapter molecules such as IRS-1 and IRS-2. Both IRS-1 and IRS-2 contain multiple tyrosine residues that, when phosphorylated through cytokine stimulation, serve as docking sites for p85 (32, 36, 40). Indeed, IRS-1 and IRS-2 have four and ten potential sites of p85 binding, respectively (40).

In addition to interactions with adapter molecules, PI-3-kinase can be brought to activated receptors through interactions with receptor-associated tyrosine kinases.

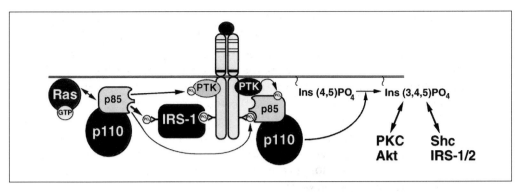

Fig. 4 The phosphoinositide-3-kinase (PI-3-kinase) pathway. Activation of PI-3-kinase occurs through the interaction of p85 with various molecules, including phosphorylated receptors, kinases, and adapters, as well as activated Ras. PI-3-kinase catalyses the production of phosphoinositides [Ins(3,4,5)PO$_4$] that activate kinases, such as protein kinase C (PKC) and Akt, and interacts with signalling molecules, such as Shc and IRS-1/2.

Such an interaction occurs between p85 and the Src-family kinase Fyn associated with the IL-2Rβ receptor (59, 60). This interaction does not require the SH2 domains of p85 but occurs through the binding of the SH3 domain of Fyn to the Pro-rich regions of p85. The p85 subunit also may be linked to receptor-associated kinases indirectly through the adapter molecule Cbl, which interacts with PI-3-kinase upon stimulation with IL-3 and also interacts with receptor-associated kinases such as Fyn and Tyk2 (61–63).

Once recruited to activated receptor complexes, PI-3-kinase is activated by several mechanisms. The direct interaction of the SH2 domains of p85 with phosphorylated receptors or adapter molecules can induce a conformational change resulting in the increased catalytic activity of the p110 subunit (64). PI-3-kinase activation can also result from the phosphorylation of p85 by receptor-associated tyrosine kinases (Fig. 4). In particular, Src-family kinases phosphorylate p85 and increase the catalytic activity of p110 (65). Although it has not been shown directly, it is likely that JAK kinase activation leads to p85 phosphorylation since engagement of different cytokine receptors that associate with JAK kinases induce the tyrosine phosphorylation of the p85 subunit (48). Activation of PI-3-kinase also may occur from its interaction with activated Ras–GTP molecules in a situation analogous to the activation of Raf-1 by Ras–GTP (66) (Fig. 4). It is not known whether this interaction with Ras–GTP activates PI-3-kinase directly or whether this interaction tethers PI-3-kinase near activated receptors and associated kinases, resulting in the activation of PI-3-kinase by these molecules. Nevertheless, this interaction represents an intersection between the Ras/MAPK and PI-3-kinase signalling pathways and may represent a common point of regulation of these pathways.

As the biochemical mechanisms that result in PI-3-kinase activation have been elucidated, the cellular events triggered by PI-3-kinase activation have been more clearly defined. A critical observation was that the phosphorylated lipids resulting from the activation of PI-3-kinase act as second messengers that activate a number of cellular kinases by interacting directly with these proteins. In particular, PtdIns(3,4,5)P$_3$ and PtdIns(3,4)P$_2$ have been shown to act as potent activators of Ca^{2+}-independent forms of protein kinase C (isozymes δ, ε, and η), which phosphorylate and alter the activity of numerous cellular substrates (67). Another important example of kinase activation by lipid intermediates is the activation by PtdIns(3,4)P$_2$ of the Akt kinase (also known as protein kinase B), a serine/threonine kinase that plays a key role in cell survival (68, 69). Akt contains a PH domain that is critical to its activation and is homologous to the PH domains found in other signalling molecules such as IRS-1 and IRS-2 (70). Interestingly, Akt activation has been correlated with the ability of its PH domain to bind PtdIns(3,4)P$_2$, which leads to the suggestion that the PH domain mediates kinase activation through its interaction with PtdIns(3,4)P$_2$.

The interaction of phosphoinositides with signalling molecules may involve SH2 and PTB domains as well as PH domains (39, 71). The ability of SH2, PTB, and PH domains to interact with phosphoinositide lipids raises the intriguing possibility that lipid second messengers, in addition to directly activating the enzymatic activity of kinases, act as membrane docking sites for signalling molecules, thus localizing these molecules at the cell surface near activated receptor complexes and membrane-associated Ras (39, 72). Thus, the binding of phosphoinositides by SH2, PTB, and PH domains in adapter molecules such as SHC, IRS-1, and IRS-2, which regulate Ras activation, represents another point of intersection between the PI-3-kinase and Ras/MAPK pathways. Indeed, activation of SHC phosphorylation, a critical step leading to Grb2/SOS interaction and Ras activation, has been shown to be require both the lipid binding capability and the phosphotyrosine binding function of the PTB domain of SHC (71). Clearly, the PI-3-kinase pathway impacts numerous cellular signalling pathways and processes through the production of phosphoinositide second messengers. However, it is likely the full extent to which this pathway impacts these processes remains to be fully determined.

3.3 Pathways leading to gene activation

Although signalling through the Ras/MAPK and PI-3-kinase pathways can ultimately result in the activation of gene expression, these pathways are generally associated with mitogenic responses induced by cytokines and other stimuli. However, one of the well-documented biological responses to cytokines stimulation is the rapid and specific induction of gene transcription. This process is mediated by a series of latent cytoplasmic transcription factors that are activated by kinases associated with cytokine receptors. Once activated, these factors translocate directly to the nucleus where they bind unique DNA sequences in the promoter regions of cytokine-responsive genes, thus stimulating the expression of these genes.

3.3.1 Signal transducers and activators of transcription

One of the most striking recent developments in cytokine signalling has been the identification and characterization of the signal transducers and activators of transcription (STAT). One or more STAT molecules have been shown to be activated by all members of the cytokine receptor superfamily. Since this family of receptors also activates JAK tyrosine kinases, the STAT activation pathway is often referred to as the JAK–STAT pathway. This is somewhat of a misnomer as STAT activation is not the sole function of JAK kinases. Nevertheless, elegant experiments utilizing mutant cell lines that lack specific JAK kinases have shown that JAK activation is required for STAT activation (73).

Thus far, seven genes encoding structurally related STAT molecules have been identified (Table 1). The STAT-1 and STAT-2 molecules were the first to be identified as transcription factors rapidly activated by interferon (IFN)-α/β stimulation, which, along with a protein termed p48, are required for the expression of IFN-responsive genes (74). Additional STAT genes were identified based on sequence homology to STAT-1 and STAT-2 and by purification of activated transcription complexes. These include STAT-3, STAT-4, the highly related STAT-5A and STAT-5B, and STAT-6. Although these STAT molecules are the only ones yet to be identified, multiple splice variants of these proteins have been characterized that may have unique functions. Interestingly, a STAT-like molecules has been identified in *D. melanogaster*, indicating the evolutionary conservation of this signalling pathway (75).

The conserved STAT genes are likely to have resulted from the duplication of a primordial gene. This concept is supported by the fact that the STAT gene loci are tightly linked on mouse chromosomes 1, 10, and 11 (76) (Table 1). Also, the overall structure of STAT molecules are highly conserved (Fig. 2C). STATs contain central SH3 and SH2 domains that are critical for STAT activation and protein–protein interaction. Flanking the SH3 domain is a DNA binding domain and a conserved N-terminal domain that is required for STAT function. Small deletions in this N-

Table 1 Characteristics of signal transducers and activators of transcription (STAT) molecules

	Activating cytokines	Chromosomal localization	Heterodimer formation	Consensus DNA binding site	Phenotype of STAT knock-outs
STAT1	IFN-α, -γ, IL-6, IL-10, growth hormone (EGF, PDGF, CSF-1, angiotensin)	Mouse: Ch.1 Human: Ch.2	STAT2, STAT3	TTCC (G/c) GGAA	Impaired innate immune response to viral and bacterial infections
STAT2	IFN-α	Mouse: Ch. 10 Human: Ch. 17	STAT1	Binds only as heterodimer	Embryonic lethal
STAT3	IFN-α, IL-2, IL-6, IL-10, IL-12	Mouse: Ch. 11 Human: Ch. 12	STAT1	TTCC (G=C) GGAA	Embryonic lethal
STAT4	IL-12, IFN-α	Mouse: Ch. 1 Human: Ch. 2	ND	TTCC (G/c) GGAA	Deficient in Th1 cells
STAT5A} STAT5B}	IL-2, IL-7, IL-15, prolactin	Mouse: Ch. 11 Human: Ch. 12	STAT5A/5B	TTCC (A>T) GGAA	5A–/–: breast tissue, lactation defective 5B–/–: growth defect
STAT6	IL-4, IL-13	Mouse: Ch. 10 Human: Ch. 17	ND	TTCC (A>T, N) GGAA	Deficient in Th2 cells

terminal region block STAT activation. The divergent C-terminal region has been implicated in transcriptional activation. Indeed, a splice variant of STAT-1, termed STAT-1β, has a 38 amino acid deletion in this C-terminal region and does not activate transcription in contrast to the full-length protein (termed STAT-1α) (77).

The structural conservation of STAT molecules also is reflected in the conserved mechanism by which these molecules are activated by cytokine receptors (Fig. 5). Cytokine receptor engagement results in the activation of receptor-associated JAK kinases and phosphorylation of specific tyrosine residues in the receptor cytoplasmic tail. The SH2 domain of the STAT molecule then binds to the phosphorylated receptor, enabling the activated kinases to phosphorylate the STAT at a highly conserved C-terminal tyrosine residue near amino acid 700 (Fig. 2C). Once phosphorylated, the

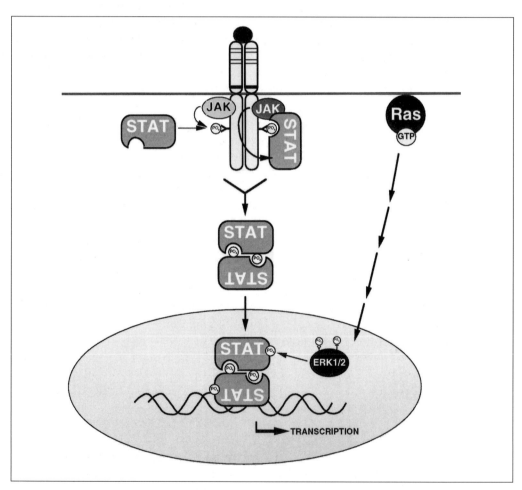

Fig. 5 The STAT activation pathway. Cytokine binding results in the activation of associated JAK kinases and receptor phosphorylation. STAT molecules then bind phosphorylated receptors, become phosphorylated, dimerize, and translocate the nucleus. Transcriptional activation by STATs may require the serine phosphorylation of STAT molecules by ERK kinases in the Ras/MAPK pathway.

STAT molecule disengages from the receptor and forms homodimers through interaction of the SH2 domain with the C-terminal phosphotyrosine residue of a second STAT molecule. When different STAT molecules are activated by a single receptor, heterodimer formation can occur (Table 1). STAT-2 in particular does not bind DNA unless it is in a heterodimer configuration.

The dimerized STATs translocate to the nucleus where they bind to specific DNA motifs in the promoters of responsive genes (Fig. 5). The DNA motifs bound by different STATs bear remarkable similarity to each other and reflect a dyad symmetry (Table 1). The exact mechanism by which STATs activate transcription is still being determined. It is likely that STAT molecules themselves activate the basic transcriptional machinery but it also has been shown that different STAT molecules form complexes with other well-characterized transcription factors such as c-Jun, SP1, and C/EBPα and thus may activate transcription through co-operative interaction with these factors. Alternatives to this general mechanism of STAT activation have been described. In one case, activation of STAT-1 by IFN-α has been shown to occur through its interaction with receptor-bound phosphorylated STAT-2, rather than by interacting with the phosphorylated receptor itself (78). In another case, activation of STAT-5 by the gp130 subunit of the IL-6 receptor does not require receptor phosphorylation. Instead, STAT-5 molecules bind phosphotyrosine residues in active JAK2 kinase molecules associated with the gp130 cytoplasmic tail, resulting in the phosphorylation and dimerization of STAT-5 (79).

While multiple cytokines stimulate the activation of STAT-1, STAT-3, and STAT-5, a very specific set of cytokines lead to the activation of STAT-2, STAT-4, and STAT-6 (Table 1). The specificity of STAT activation may be attained on several levels. The primary control of STAT activation is through the specific recognition of tyrosine phosphorylated receptor sequence by STAT SH2 domains. This has been elegantly demonstrated in experiments where short peptide sequences required for STAT-3 activation by the gp130 subunit of the IL-6 receptor were transferred to the erythropoietin receptor, conferring on this receptor the ability to activate STAT-3 (80). The specificity of STAT SH2 domains was demonstrated in experiments where the SH2 domain of STAT-2 was replaced with the SH2 domain of STAT-1 (81). The resulting chimeric STAT-2 molecule was then able to be activated by the IFN-γ receptor, which normally activates STAT-1 but not STAT-2. The specificity of STAT activation also may be controlled by the limited expression of cytokine receptors that activate the same STAT molecules, such that different receptors activating the same STAT are not expressed in the same cell.

It is also possible that cytokines activate additional signalling pathways that differentially contribute to the STAT activation pathways. An important example is that activation of the Ras/MAPK pathway is required for the full function of the STAT-1α and STAT-3 molecules. In particular, a serine residue in the C-terminal region of STAT-1α was phosphorylated in response to activation of the Ras/MAPK pathway, presumably through the action of MAPKs such as ERK1/2, and mutation of this serine resulted in a threefold reduction in STAT-induced transcriptional activation (82). IFN-α/β are not efficient activators of the Ras/MAPK pathway.

However, activation of the MAPK isoform ERK2 occurs through the direct inter-action of ERK2 with the α subunit of the IFN-α/β receptor (83). ERK2 and STAT-1α were also demonstrated to directly interact upon IFN-β stimulation. Thus, the full activation of STAT-1α can be achieved through the direct interaction and activation of a MAPK by the activating cytokine receptor.

In additional studies, serine phosphorylation of STAT-3 was shown to be required for optimal DNA binding (84). This is in contrast to the findings with STAT-1α, in which serine phosphorylation did not seem to alter DNA binding, but this may represent another level of control of STAT activation by kinase cascades such as the Ras/MAPK cascade. Interestingly, STAT-3 also has been demonstrated to act as an adapter molecule linking the α subunit of the IFN-α/β receptor to the p85 subunit of PI-3-kinase, leading to the activation of PI-3-kinase (85). Taken together, these results indicate points at which the STAT activation pathway intersects with additional signalling pathways, thus leading to a higher level of specificity in STAT activation and a co-ordination of cytokine-stimulated intracellular signals.

The critical importance of STAT activation *in vivo* has been demonstrated in studies on mice in which STAT genes have been knocked-out through homologous recombination in embryonic stem cells. Significantly, these studies have suggested that while many stimuli can activate certain STATs *in vitro*, the activation and ulti-mate function of STATs *in vivo* can be very specific (Table 1). The is best exemplified in mice lacking STAT-1. In spite of the fact that STAT-1 is activated by multiple stimuli, mice lacking STAT-1 only appear to have a deficiency in their innate im-munity to viral and bacterial infection that is mediated by interferons (86) (Table 1). No deficiency was observed that could be attributed to a lack of signalling by other cytokines that activate STAT-1 *in vitro*. It remains to be determined whether additional phenotypes will result from interbreeding mice deficient in the more than one STAT gene. It is interesting to speculate that additional functions that result from the activation of multiple STAT pathways will be uncovered in these mice.

3.3.2 Additional cytokine-activated transcription factors

The expression of many genes is increased in inflammatory and immune responses. The control of the expression of these genes may involve a number of transcription factors including STATs and transcription factors activated by the Ras/MAPK pathway. However, two transcription factors in particular, termed NF-IL6 and NFκB, have been implicated in the cytokine-stimulated expression of genes important in inflammatory, immune, and acute phase responses. Like STATs, these transcription factors are activated from a latent cytoplasmic state and translocate to the nucleus where they activate specific genes.

3.3.3 NF-IL6

NF-IL6 was first identified as the transcription factor that bound the IL-1-responsive element in the IL-6 gene promoter and is involved in the activation of a number of genes including IL-1, IL-4, IL-8, tumour necrosis factor (TNF)-α, and G-CSF (87). It is a member of a family of related factors that includes C/EBP, NF-IL6β, Ig/EBP, and CHOP. Members of this family contain C-terminal DNA binding domains composed of basic amino acid residues and a leucine zipper dimerization motif. NF-IL6 is expressed widely and is induced by IL-1, TNF, IL-6, and lipopolysaccharide (LPS) stimulation. Activation of latent NF-IL6 requires serine/threonine phosphorylation, which is catalysed by a Ras-dependent MAPK and results in the ability of this factor to activate transcription of responsive genes.

3.3.4 NFκB

One of the best studied of all the transcription factors, NFκB, was first identified as an inducible factor binding to the kappa immunoglobulin intronic enhancer and is involved in the regulation of many genes including IL-6, IL-8, TNF, lymphotoxin, IFN-β, and the IL-2 receptor. The most prevalent form of activated NFκB is composed of two subunits, p50 and p65 (RelA), which are members of a family of proteins related to the c-*rel* proto-oncogene. The p50/p65 heterodimer exists in an inactive form in the cytoplasm of most cells complexed to a protein termed IκB-α, which itself is a member of a family of IκB proteins (88). The inhibitory function of IκB proteins involves the masking of nuclear localization signal sequences within the NFκB heterodimer. A number of stimuli, including those induced by the cytokines IL-1 and TNF-α, cause the dissociation of IκB-α from the NFκB complex and result in the nuclear translocation of NFκB and subsequent activation of responsive genes. This dissociation is initiated by the phosphorylation of IκB-α on two serine residues by two recently identified kinases termed IKK-α and IKK-β (for *IkB* Kinases α and β) (89, 90). Once phosphorylated, IκB-α is rapidly linked to multiple chains of the 76 amino acid protein ubiquitin. This process targets the IκB-α protein for degradation by multisubunit complexes termed proteosomes. Degradation of IκB-α results in the release and nuclear translocation of NFκB. As discussed in the next section, this activation of NFκB also appears to play a critical role in the prevention of cytokine-stimulated cell death by apoptosis.

3.4 Pathways leading to cell death

The ability of many cytokines to activate the Ras/MAPK and the PI-3-kinase pathways results in not only cellular proliferation but also in the prevention of cell death by apoptosis (91). In contrast to these cytokines, a set of secreted factors are capable of activating specific signalling pathways that result in apoptosis. These secreted factors are typified by the cytokine tumour necrosis factor α (TNF-α) which is produced by infiltrating leukocytes during inflammatory responses. TNF-α is unique in this set of 'death factors' in that it also induces activation of NFκB, which counteracts the apoptotic signals and results in cellular proliferation. Thus, a fine balance exists in TNF-α stimulated cells because of the concomitant activation of apoptotic and NFκB signalling pathways. The outcome of TNF-α stimulation, be it cell survival or death, may depend on a number of factors.

The secreted factors related to TNF-α include another cytokine, lymphotoxin, as well as the ligands for the Fas, CD40, as well CD30 receptors (see Chapter 1). The soluble, active forms of all these factors exist as trimers that induce receptor trimerization upon binding. Two forms of the TNF-α receptor (TNFR) exist including the 55 kDa TNFR1, which is used predominantly, and the 75 kDa TNFR2, which is important in thymocyte development. The cytoplasmic domains of these receptors contain a specific interaction sequence, termed a death domain, that is bound by a series of signalling molecules that also contain death domains. These death domain–death domain interactions are integral to the TNF-α-activated apoptotic pathway and a number of additional molecules containing death domains have been identified (92). The precise order of interaction with the TNFR and the resultant function of the death domain signalling molecules is still being delineated (92). However, it is clear that the end-result of these interactions is the activation of two cytoplasmic proteins, the aforementioned NFκB and caspase-8, a protease whose activation leads to apoptosis.

Caspase-8 (for *cys*-aspase-8) is a cysteine protease that specifically cleaves substrates after aspartic acid residues and is related to a number of proteases including the IL-1β converting enzyme (ICE). Caspase-8 exists as a inactive precursor, or zymogen, that is activated by autocatalytic cleavage after its interaction with death domain molecules bound to cross-linked TNFR. Activation of caspase-8 leads to the activation other ICE-like proteases through a cascade of proteolytic cleavages analogous to the kinase cascade of the Ras/MAPK pathway. Ultimately, cleavage of cellular substrates by these proteases leads to the breakdown of cellular structure and the degradation of chromosomal DNA characteristic of apoptosis.

In addition to activating caspase-8, TNFR engagement also activates the serine/threonine kinases, RIP and NIK, that contribute to the activation of NFκB. In particular, NIK has been shown to interact with and activate the IKK-α and IKK-β kinases that are critical to inactivating the cytoplasmic NFκB inhibitor, IkB-α (89, 90). Blocking the activation NFκB, either by knocking-out NFκB genes or expressing inhibitors of NFκB activation, greatly enhances the ability of TNF-α to induce apoptosis (93–95). Thus, it has been proposed that NFκB induces a gene or set of genes that encode proteins with anti-apoptotic functions. However, these proteins and the mechanism by which they counter the protease cascade activated by TNF-α remain to be characterized.

It also remains to be determined how the balance between the apoptotic and anti-apoptotic pathways is maintained in TNF-α-stimulated cells. The outcome in some cases may depend on the specific cell stimulated by TNF-α as evidenced by the fact that some tumours regress upon TNF-α treatment, suggesting a predominance of apoptosis, while others do not. The outcome may also reflect the general state of the cell. A proliferating cell is likely to have a number of signalling pathways functioning simultaneously and these pathways may push the balance of TNF-α stimulation away from apoptosis. In this regard, activation of both the Ras/MAPK and PI-3-kinase pathways have been shown to specifically inhibit apoptosis in other systems (91). Controlling the TNF-α signalling balance may be very important clinically. For instance, if NFκB activation could be inhibited, the efficacy of TNF-α as a cancer treatment could be greatly enhanced by tipping the balance toward apoptosis in TNF-α-treated tumours.

4. Modulation of signal transduction pathways

One of the most important recent developments in the cytokine field has been the delineation of mechanisms that lead to the modulation or negative regulation of cytokine signalling. Although some of these mechanisms are involved in the control of a number of signalling pathways, others appear to be specific to the regulation of cytokine signalling pathways.

4.1 Negative regulation of kinase activation

The inactivation of receptor-associated tyrosine kinases is a critical regulatory mechanism in all signalling pathways and is generally considered to occur through the action of phosphatases. However, the components of a novel negative regulatory pathway involved in the specific inactivation of JAK kinases have recently been described (96–98). In this pathway, cytokine stimulation results in the expression of a series of related SH2-domain proteins that have been termed the SOCS-1, 2, 3 (suppressors of cytokine signalling), JAB (JAK binding), CIS (cytokine-induced SH2), and SSI-1 (STAT-induced STAT-inhibitor) proteins (96–98) (Fig. 6). Once expressed, these proteins bind to phosphorylated JAK kinases and inhibit their function. The proteins appear to be specific for JAK kinases since STAT activation is inhibited while the phosphorylation of proteins by other Src-family kinases is unaffected. Expression of the genes encoding these proteins appears to require STAT activation. Thus this pathway represents a classical negative feedback loop where induction of the JAK–STAT pathway results in its eventual inactivation through the induction of SOCS/JAB/CIS/SSI expression. Then, as inhibitor expression diminishes from in-activation of the JAK–STAT pathway, the cell is returned to a state favourable to the subsequent reactivation of the JAK–STAT pathway.

Like the JAK kinases, kinases in the Ras/MAPK pathway also appear to be specifically regulated. In particular, a family of genes encoding MAPK-specific phosphatases has been identified (99). These phosphatases catalyse the hydrolysis of phosphate from phosphotyrosine and serine residues in activated MAPK, resulting in its inactivation. Interestingly, the expression of the phosphatases, MKP-1 and MKP-2, is induced via activation of the Ras/MAPK pathway (99) (Fig. 6). Thus, activation of the Ras/MAPK pathway can result in the expression of specific phosphatases that attenuate the function of MAPK. This results in a negative feedback loop reminiscent of that seen in the JAK–STAT pathway.

4.2 Negative regulation of the Ras/MAPK pathway

In addition to the specific regulation of MAPK activation through the action of MAPK phosphatases, recent findings have suggested that specific negative

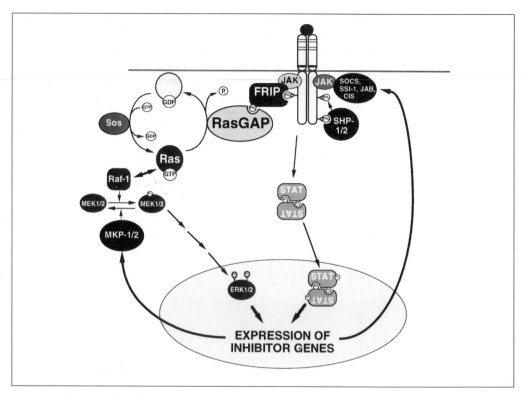

Fig. 6 Negative regulation of different cytokine signalling pathways. Negative regulatory molecules are shown in black. STATs activate genes encoding inhibitory molecules (SOCS, SSI-1, JAB, CIS) that inhibit JAK function. Activation of the Ras/MAPK pathway leads to the expression of the MPK-1/2 phosphatases that inactivate MEKs in the Ras/MAPK pathway. FRIP may link RasGAP to activated receptors resulting in Ras inactivation. The SHP-1/2 phosphatases act by dephosphorylating signalling molecules such as cytokine receptors.

regulation of the Ras/MAPK pathway may occur at the point of Ras activation. Like Ras activation, this regulation involves the function of specific adapter molecules. Negative regulation of Ras activation has been shown to be an important mechanism leading to a non-responsiveness in T cells. This non-responsive state, termed anergy, is thought to be important in inhibiting proliferation of autoreactive T cells, thus preventing autoimmunity. Negative regulation of Ras activation in anergic T cells has recently been shown to result, in part, from the hyperphosphorylation of the Cbl adapter (100). Cbl hyperphosphorylation leads to its interaction with the Crk/C3G guanine-nucleotide exchange complex that activates the Ras-related protein Rap1, but not Ras. Rap1, like Ras, is a small GTPase but Rap1 is not associated with the plasma membrane and has been shown to inhibit activation of the Ras/MAPK pathway presumably by acting as a cytoplasmic competitor for Ras effector molecules. Thus, Cbl hyperphosphorylation can result in Rap1 hyperactivation in anergic T cells and thus prevent T cell proliferation by blocking the activation of the Ras/MAPK pathway. Since mitogenic cytokines such as IL-3 induce the phosphorylation of Cbl and its interaction with Crk, it is possible that Cbl may play a role in modulating Ras/MAPK pathway activation induced by cytokines (101).

A number of adapter molecules mediate the interaction of Sos with activated cytokine receptor complexes and result in the accumulation of activated Ras–GTP. Two recently identified adapter proteins, p62^{dok} (Dok) and FRIP, may contribute to the process of Ras regulation by interacting with RasGAP, which acts in opposition to Sos. Dok (downstream of kinases) is a phosphoprotein that has long been observed to interact with RasGAP (102, 103) (Fig. 2B). Although its function has yet to be clearly delineated, Dok has domains common to adapter molecules such as N-terminal PH and PTB domains, and a C-terminal region that contains consensus binding sites for the N-terminal SH2 domain of RasGAP (102, 103). FRIP (interleukin-four receptor interacting protein) is highly homologous to Dok but was cloned based on the ability of its PTB domain to bind to a tyrosine phosphorylated motif in the IL-4 receptor and has been demonstrated to be phosphorylated in response to different cytokine including IL-2, IL-3, and IL-4 (K. Nelms, in press). In contrast to Dok, which is expressed in a wide variety of tissues, FRIP expression is limited to haematopoietic cells and is expressed at particularly high levels in T cells. Cytokine-induced phosphorylation of FRIP increases its interaction with RasGAP suggesting that FRIP interacts with phosphorylated cytokine receptors and becomes phosphorylated by receptor-associated tyrosine kinases, which then enables it to interact with RasGAP (Fig. 6). Thus, FRIP may link RasGAP to activated receptor complexes, resulting in the increased hydrolysis of Ras–GTP and inactivation of the Ras/MAPK pathway.

4.3   Regulation of signal transduction by phosphatases

The dephosphorylation of tyrosine phosphorylated proteins is a critical process in the modulation of cytokine-stimulated signal transduction and the proteins involved in this process are now beginning to be understood. Three phosphatases in particular have been shown to be important in regulating a number of signalling pathways.

SH2-containing phosphatase-1 (SHP-1; previously know as HCP, SHPTP1, PTP1C) and SH2-containing phosphatase-2 (SHP-2; previously known as Syp, SHPTP2) are highly related phosphatases that contain tandem, N-terminal SH2 domains (104) (Fig. 6). SHP-1 and SHP-2 are known to interact directly with phosphorylated receptors via their SH2 domains. The specific sequences recognized by the SHP-1/2 have been termed immunoregulatory tyrosine-based inhibitory motifs (ITIM) because of their presence in a number of receptors associated with immune cell function such as the FcγRIIb1 immunoglobulin receptor. Interaction with these motifs is thought to lead to SHP activation and bring each phosphatase into contact with specific substrates such as receptor-associated kinases.

SHP-1 is expressed primarily in haematopoietic cells where it is important in the regulation of signal transduction, as demonstrated in studies performed on the *motheaten* mouse. Mice homozygous for the *motheaten* allele have a number of haematopoietic abnormalities, including hyperproliferation and increased phosphorylation of protein substrates such as JAK1 (105). These abnormalities result from a mutation that prevents SHP-1 expression. These studies also have suggested that SHP-1 action is specific to certain signalling pathways, since activation of a different JAK kinase, Tyk2, by IFN-α stimulation was normal in *motheaten* and control mice. It is not clear what other phosphatase(s) function in *motheaten* mice and regulate pathways such as those induced by IFN-α, but SHP-2 is a likely candidate. Although the specific role played by SHP-1/2 in cytokine signalling is not yet defined, a sequence resembling an ITIM motif is evident in the cytoplasmic domain of the IL-4 receptor α chain. Similar motifs exist in other receptors and thus may indicate a general role of these phosphatases in cytokine signal modulation.

A third phosphatase, termed SHIP (*SH2* containing inositol-5′-phosphatase), with specificity for the 3′ and 5′ phosphoinositides of $PtdIns(3,4,5)P_3$, appears to regulate the PI-3-kinase pathway by dephosphorylating the substrates of this enzyme (104). This phosphatase activity, however, does not necessarily result in the negative regulation of the PI-3-kinase pathway. In particular, the formation of $PtdIns(3,4,)P_2$, a critical activator of the anti-apoptotic kinase Akt, results from the dephosphorylation of $PtdIns(3,4,5)P_3$ by SHIP (68). As with other signalling molecules, activation of SHIP may involve its interaction with adapter molecules. Indeed, a number of cytokines have been shown to induce the interaction of SHIP with the phosphorylated adapter molecules SHC (106, 107). This result also suggests that SHIP may be involved in the regulation of the Ras/MAPK pathway. The mechanisms that lead to activation of phosphatases by cytokines are only now being elucidated, and determining the specific role that SHIP and other phosphatases play in cytokine signalling will be an area of intense focus in the near future.

5. Conclusion

Multiple and distinct signalling pathways are activated simultaneously by the engagement of individual cytokine receptors. The best understood of the cytokine signalling pathways have been outlined in this chapter. Since these pathways utilize unique molecular components, it is tempting to view them as distinct and independent pathways that lead to specific cellular responses. However, an important conclusion that can be made from the analysis of these individual pathways is that they do not act autonomously but rather intersect and influence each other at different points, forming an intracellular signalling network that can initiate multiple intracellular events that lead to specific cellular responses. Thus, the array of cellular responses elicited by a cytokine is not only a reflection of the signalling pathways activated by the cytokine but of the degree of interaction between these different pathways.

Thus, an important challenge for the future is to understand in detail how different cytokine signalling pathways act in concert to elicit various cellular responses. This will require a detailed knowledge of pathways activated by each cytokine as well as a determination of the degree to which each individual signalling pathway contributes to a particular cellular response. Ultimately, as yet uncharacterized signalling pathways may need to be fully delineated before this is possible. It is probable that important tools for defining additional signalling pathways will be the growing database of expressed DNA sequences that may be scanned for sequences homologous to the already characterized signalling molecules, as well as highly sensitive sequencing techniques that enable sequence to be obtained from minute amounts of protein. Such technical advances, in addition to the analysis of cells and mice genetically engineered to be deficient in specific signalling molecules, will certainly lead to a more complete understanding of the complex interactions and biological functions of the intracellular signalling pathways activated by cytokines.

References

1. Miyajima, A., Kitamura, T., Harada, N., Yokota, T., and Arai, K. (1992) Cytokine receptors and signal transduction. *Annu. Rev. Immunol.*, **10**, 295.
2. Taniguchi, T. (1995) Cytokine signaling through nonreceptor protein tyrosine kinases. *Science*, **268**, 251.
3. Ryan, J. J., McReynolds, L. J., Keegan, A., Wang, L. H., Garfein, E., Rothman, P., *et al.* (1996) Growth and gene expression are predominantly controlled by distinct regions of the human IL-4 receptor. *Immunity*, **4**, 123.
4. Ihle, J. N. (1995) The Janus protein tyrosine kinase family and its role in cytokine signaling. *Adv. Immunol.*, **60**, 1.
5. Duronio, V., Welham, M. J., Abraham, S., Dryden, P., and Schrader, J. W. (1992) p21ras activation via hemopoietin receptors and c-*kit* requires tyrosine kinase activity but not tyrosine phosphorylation of p21ras GTPase-activating protein. *Proc. Natl. Acad. Sci. USA*, **89**, 1587.
6. Welham, M. J., Duronio, V., Sanghera, J. S., Pelech, S. L., and Schrader, J. W. (1992) Multiple hemopoietic growth factors stimulate activation of mitogen-activated protein kinase family members. *J. Immunol.*, **149**, 1683.
7. Welham, M. J., Duronio, V., Leslie, K. B., Bowtell, D., and Schrader, J. W. (1994) Multiple hemopoietins, with the exception of interleukin-4, induce modification of Shc and mSos1, but not their translocation. *J. Biol. Chem.*, **269**, 21165.
8. Crowley, M. T., Harmer, S. L., and DeFranco, A. L. (1996) Activation-induced association of a 145-kDa tyrosine-phosphorylated protein with Shc and Syk in B lymphocytes and macrophages. *J. Biol. Chem.*, **271**, 1145.
9. Wery, S., Letourneur, M., Bertoglio, J., and Pierre, J. (1996) Interleukin-4 induces activation of mitogen-activated protein kinase and phosphorylation of shc in human keratinocytes. *J. Biol. Chem.*, **271**, 8529.
10. Marshall, C. J. (1994) MAP kinase kinase kinase, MAP kinase kinase and MAP kinase. *Curr. Opin. Genet. Dev.*, **4**, 82.
11. Marshall, C. J. (1996) Ras effectors. *Curr. Opin. Cell. Biol.*, **8**, 197.
12. Gotoh, T., Hattori, S., Nakamura, S., *et al.* (1995) Identification of Rap1 as a target for the Crk SH3 domain-binding guanine nucleotide-releasing factor C3G. *Mol. Cell. Biol.*, **15**, 6746.
13. Boguski, M. S. and McCormick, F. (1993) Proteins regulating Ras and its relatives. *Nature*, **366**, 643.
14. Hancock, J. F., Paterson, H., and Marshall, C. J. (1990) A polybasic domain or

palmitoylation is required in addition to the CAAX motif to localize p21ras to the plasma membrane. *Cell*, **63**, 133.

15. Hall, A. and Self, A. J. (1986) The effect of Mg^{2+} on the guanine nucleotide exchange rat of p21 N-ras. *J. Biol. Chem.*, **261**, 10963.

16. Trahey, M. and McCormick, F. (1987) A cytoplasmic protein stimulates normal N-ras p21 GTPase, but does not affect oncogenic mutants. *Science*, **238**, 542.

17. Pawson, T. (1995) Protein modules and signalling networks. *Nature*, **373**, 573.

18. Henkemeyer, M., Rossi, D. J., Holmyard, D. P., Puri, M. C., Mbamalu, G., Harpal, K., *et al.* (1995) Vascular system defects and neuronal apoptosis in mice lacking ras GTPase-activating protein. *Nature*, **377**, 695.

19. Jacks, T., Shih, T. S., Schmitt, E. M., Bronson, R. T., Bernards, A., and Weinberg, R. A. (1994) Tumour predisposition in mice heterozygous for a targeted mutation in Nf1. *Nature Genet.*, **7**, 353.

20. Downward, J. (1996) Control of ras activation. *Cancer Surv.*, **27**, 87.

21. Bowtell, D. D. and Langdon, W. Y. (1995) The protein product of the *c-cbl* oncogene rapidly complexes with the EGF receptor and is tyrosine phosphorylated following EGF stimulation. *Oncogene*, **11**, 1561.

22. Chardin, P., Camonis, J. H., Gale, N. W., van Aelst, L., Schlessinger, J., Wigler, M. H., *et al.* (1993) Human Sos1: a guanine nucleotide exchange factor for Ras that binds to GRB2. *Science*, **260**, 1338.

23. Gulbins, E., Coggeshall, K. M., Baier, G., Katzav, S., Burn, P., and Altman, A. (1993) Tyrosine kinase-stimulated guanine nucleotide exchange activity of Vav in T cell activation [see comments]. *Science*, **260**, 822.

24. Clark, S. G., Stern, M. J., and Horvitz, H. R. (1992) *C. elegans* cell-signalling gene sem-5 encodes a protein with SH2 and SH3 domains [see comments]. *Nature*, **356**, 340.

25. Olivier, J. P., Raabe, T., Henkemeyer, M., Dickson, B., Mbamalu, G., Margolis, B., *et al.* (1993) A *Drosophila* SH2–SH3 adaptor protein implicated in coupling the sevenless tyrosine kinase to an activator of Ras guanine nucleotide exchange, Sos. *Cell*, **73**, 179.

26. Simon, M. A., Dodson, G. S., and Rubin, G. M. (1993) An SH3–SH2–SH3 protein is required for p21Ras1 activation and binds to sevenless and Sos proteins *in vitro*. *Cell*, **73**, 169.

27. Songyang, Z., Shoelson, S. E., Chaudhuri, M., *et al.* (1993) SH2 domains recognize specific phosphopeptide sequences. *Cell*, **72**, 767.

28. Trub, T., Choi, W. E., Wolf, G., Ottinger, E., Chen, Y., Weiss, M., *et al.* (1995) Specificity of the PTB domain of Shc for beta turn-forming pentapeptide motifs amino-terminal to phosphotyrosine. *J. Biol. Chem.*, **270**, 18205.

29. Salcini, A. E., McGlade, J., Pelicci, G., Nicoletti, I., Pawson, T., and Pelicci, P. G. (1994) Formation of Shc–Grb2 complexes is necessary to induce neoplastic transformation by overexpression of Shc proteins. *Oncogene*, **9**, 2827.

30. Rozakis-Adcock, M., McGlade, J., Mbamalu, G., *et al.* (1992) Association of the Shc and Grb2/Sem5 SH2-containing proteins is implicated in activation of the Ras pathway by tyrosine kinases. *Nature*, **360**, 689.

31. Gotoh, N., Tojo, A., and Shibuya, M. (1996) A novel pathway from phosphorylation of tyrosine residues 239/240 of Shc, contributing to suppress apoptosis by IL-3. *EMBO J.*, **15**, 6197.

32. Sun, X. J., Rothenberg, P., Kahn, C. R., Backer, J. M., Araki, E., Wilden, P. A., *et al.* (1991) Structure of the insulin receptor substrate IRS-1 defines a unique signal transduction protein. *Nature*, **352**, 73.

33. Sun, X. J., Crimmins, D. L., Myers, M. G., Miralpeix, M., and White, M. F. (1993) Pleiotropic insulin signals are engaged by multisite phosphorylation of IRS-1. *Mol. Cell. Biol.*, **13**, 7418.
34. Wang, L. M., Keegan, A. D., Li, W., Lienhard, G. E., *et al.* (1993) Common elements in interleukin-4 and insulin signaling pathways in factor-dependent hematopoietic cells. *Proc. Natl. Acad. Sci. USA*, **90**, 4032.
35. Keegan, A. D., Nelms, K., White, M., Wang, L. M., Pierce, J. H., and Paul, W. E. (1994) An IL-4 receptor region containing an insulin receptor motif is important for IL-4-mediated IRS-1 phosphorylation and cell growth. *Cell*, **76**, 811.
36. Johnston, J. A., Wang, L. M., Hanson, E. P., Sun, X. J., White, M. F., Oakes, S. A., *et al.* (1995) Interleukins 2, 4, 7 and 15 stimulate tyrosine phosphorylation of insulin receptor substrates 1 and 2 in T cells. Potential role of JAK kinases. *J. Biol. Chem.*, **270**, 28527.
37. Zhou, M. M., Huang, B., Olejniczak, E. T., Meadows, R. P., Shuker, S. B., Miyazaki, M., *et al.* (1996) Structural basis for IL-4 receptor phosphopeptide recognition by the IRS-1 PTB domain. *Nature Struct. Biol.*, **3**, 388.
38. Eck, M. J., Dhe-Paganon, S., Trub, T., Nolte, R. T., and Shoelson, S. E. (1996) Structure of the IRS-1 PTB domain bound to the juxtamembrane region of the insulin receptor. *Cell*, **85**, 695.
39. Lemmon, M. A., Ferguson, K. M., and Schlessinger, J. (1996) PH domains: diverse sequences with a common fold recruit signaling molecules to the cell surface. *Cell*, **85**, 621.
40. Sun, X. J., Wang, L. M., Zhang, Y., Yenush, L., Myers, M. G. J., Glasheen, E., *et al.* (1995) Role of IRS-2 in insulin and cytokine signalling. *Nature*, **377**, 173.
41. Stokoe, D., Macdonald, S. G., Cadwallader, K., Symons, M., and Hancock, J. F. (1994) Activation of Raf as a result of recruitment to the plasma membrane [see comments] [published erratum appears in *Science* (1994) Dec. 16; **266**(5192), 1792]. *Science*, **264**, 1463.
42. Leevers, S. J., Paterson, H. F., and Marshall, C. J. (1994) Requirement for Ras in Raf activation is overcome by targeting Raf to the plasma membrane. *Nature*, **369**, 411.
43. Marais, R., Light, Y., Paterson, H. F., and Marshall, C. J. (1995) Ras recruits Raf-1 to the plasma membrane for activation by tyrosine phosphorylation. *EMBO J.*, **14**, 3136.
44. Turner, B. C., Tonks, N. K., Rapp, U. R., and Reed, J. C. (1993) Interleukin 2 regulates Raf-1 kinase activity through a tyrosine phosphorylation-dependent mechanism in a T-cell line. *Proc. Natl. Acad. Sci. USA*, **90**, 5544.
45. Kolch, W., Heidecker, G., Kochs, G., Hummel, R., Vahidi, H., Mischak, H., *et al.* (1993) Protein kinase C alpha activates RAF-1 by direct phosphorylation. *Nature*, **364**, 249.
46. Denhardt, D. T. (1996) Signal-transducing protein phosphorylation cascades mediated by Ras/Rho proteins in the mammalian cell: the potential for multiplex signalling. *Biochem. J.*, **318**, 729.
47. Davis, R. J. (1995) Transcriptional regulation by MAP kinases. *Mol. Reprod. Dev.*, **42**, 459.
48. Gold, M. R., Duronio, V., Saxena, S. P., Schrader, J. W., and Aebersold, R. (1994) Multiple cytokines activate phosphatidylinositol 3-kinase in hemopoietic cells. Association of the enzyme with various tyrosine-phosphorylated proteins. *J. Biol. Chem.*, **269**, 5403.
49. Augustine, J. A., Sutor, S. L., and Abraham, R. T. (1991) Interleukin 2- and poliomavirus middle T antigen-induced modification of phosphatidylinositol 3-kinase activity in activated T lymphocytes. *Mol. Cell. Biol.*, **11**, 4431.
50. Merida, I., Diez, E., and Gaulton, G. N. (1991) IL-2 binding activates a tyrosine-phosphorylated phosphatidylinositol-3-kinase. *J. Immunol.*, **147**, 2202.
51. Kapeller, R., Prasad, K. V., Janssen, O., Hou, W., Schaffhausen, B. S., Rudd, C. E., *et al.* (1994) Identification of two SH3-binding motifs in the regulatory subunit of phosphatidylinositol 3-kinase. *J. Biol. Chem.*, **269**, 1927.

52. Klippel, A., Escobedo, J. A., Hu, Q., and Williams, L. T. (1993) A region of the 85-kilodalton (kDa) subunit of phosphatidylinositol 3-kinase binds the 110-kDa catalytic subunit *in vivo*. *Mol. Cell. Biol.*, **13**, 5560.

53. Dhand, R., Hara, K., Hiles, I., Bax, B., Gout, I., Panayotou, G., *et al.* (1994) PI 3-kinase: structural and functional analysis of intersubunit interactions. *EMBO J.*, **13**, 511.

54. Dhand, R., Hiles, I., Panayotou, G., *et al.* (1994) PI 3-kinase is a dual specificity enzyme: autoregulation by an intrinsic protein-serine kinase activity. *EMBO J.*, **13**, 522.

55. Lam, K., Carpenter, C. L., Ruderman, N. B., Friel, J. C., and Kelly, K. L. (1994) The phosphatidylinositol 3-kinase serine kinase phosphorylates IRS-1. Stimulation by insulin and inhibition by Wortmannin. *J. Biol. Chem.*, **269**, 20648.

56. Stephens, L. R., Jackson, T. R., and Hawkins, P. T. (1993) Agonist-stimulated synthesis of phosphatidylinositol(3,4,5)-trisphosphate: a new intracellular signalling system? *Biochim. Biophys. Acta*, **1179**, 27.

57. Auger, K. R., Serunian, L. A., Soltoff, S. P., Libby, P., and Cantley, L. C. (1989) PDGF-dependent tyrosine phosphorylation stimulates production of novel polyphosphoinositides in intact cells. *Cell*, **57**, 167.

58. Truitt, K. E., Mills, G. B., Turck, C. W., and Imboden, J. B. (1994) SH2-dependent association of phosphatidylinositol 3'-kinase 85-kDa regulatory subunit with the interleukin-2 receptor beta chain. *J. Biol. Chem.*, **269**, 5937.

59. Pleiman, C. M., Hertz, W. M., and Cambier, J. C. (1994) Activation of phosphatidyl-inositol-3' kinase by Src-family kinase SH3 binding to the p85 subunit. *Science*, **263**, 1609.

60. Prasad, K. V., Janssen, O., Kapeller, R., Raab, M., Cantley, L. C., and Rudd, C. E. (1993) Src-homology 3 domain of protein kinase p59fyn mediates binding to phosphatidylinositol 3-kinase in T cells. *Proc. Natl. Acad. Sci. USA*, **90**, 7366.

61. Anderson, S. M., Burton, E. A., and Koch, B. L. (1997) Phosphorylation of Cbl following stimulation with interleukin-3 and its association with Grb2, Fyn and phosphatidyl-inositol 3-kinase. *J. Biol. Chem.*, **272**, 739.

62. Uddin, S., Gardziola, C., Dangat, A., Yi, T., and Platanias, L. C. (1996) Interaction of the c-*cbl* proto-oncogene product with the Tyk-2 protein tyrosine kinase. *Biochem. Biophys. Res. Commun.*, **225**, 833.

63. Fukazawa, T., Reedquist, K. A., Trub, T., Soltoff, S., Panchamoorthy, G., Druker, B., *et al.* (1995) The SH3 domain-binding T cell tyrosyl phosphoprotein p120. Demonstration of its identity with the c-*cbl* protooncogene product and *in vivo* complexes with Fyn, Grb2 and phosphatidylinositol 3-kinase. *J. Biol. Chem.*, **270**, 19141.

64. Shoelson, S. E., Sivaraja, M., Williams, K. P., Hu, P., Schlessinger, J., and Weiss, M. A. (1993) Specific phosphopeptide binding regulates a conformational change in the PI 3-kinase SH2 domain associated with enzyme activation. *EMBO J.*, **12**, 795.

65. Ruiz-Larrea, F., Vicendo, P., Yaish, P., *et al.* (1993) Characterization of the bovine brain cytosolic phosphatidylinositol 3-kinase complex. *Biochem. J.*, **290**, 609.

66. Rodriguez-Viciana, P., Warne, P. H., Dhand, R., Vanhaesebroeck, B., Gout, I., Fry, M. J., *et al.* (1994) Phosphatidylinositol-3-OH kinase as a direct target of Ras. *Nature*, **370**, 527.

67. Toker, A., Meyer, M., Reddy, K. K., Falck, J. R., Aneja, R., Aneja, S., *et al.* (1994) Activation of protein kinase C family members by the novel polyphosphoinositides PtdIns-3,4-P2 and PtdIns-3,4,5-P3. *J. Biol. Chem.*, **269**, 32358.

68. Franke, T. F., Kaplan, D. R., Cantley, L. C., and Toker, A. (1997) Direct regulation of the Akt proto-oncogene product by phosphatidylinositol-3,4-bisphosphate. *Science*, **275**, 665.

69. Kennedy, S. G., Wagner, A. J., Conzen, S. D., Jordan, J., Bellacosa, A., Tsichlis, P. N., *et al.*

(1997) The PI 3-kinase/Akt signaling pathway delivers an anti-apoptotic signal. *Genes Dev.*, **11**, 701.

70. Klippel, A., Kavanaugh, W. M., Pot, D., and Williams, L. T. (1997) A specific product of phosphatidylinositol 3-kinase directly activates the protein kinase Akt through its pleckstrin homology domain. *Mol. Cell. Biol.*, **17**, 338.

71. Ravichandran, K. S., Zhou, M. M., Pratt, J. C., Harlan, J. E., Walk, S. F., Fesik, S. W., *et al.* (1997) Evidence for a requirement for both phospholipid and phosphotyrosine binding via the Shc phosphotyrosine-binding domain in vivo. *Mol. Cell. Biol.*, **17**, 5540.

72. Lemmon, M. A., Ferguson, K. M., Sigler, P. B., and Schlessinger, J. (1995) Specific and high-affinity binding of inositol phosphates to an isolated pleckstrin homology domain. *Proc. Natl. Acad. Sci. USA*, **92**, 10472.

73. Velazquez, L., Fellous, M., Stark, G. R., and Pellegrini, S. (1992) A protein tyrosine kinase in the interferon alpha/beta signaling pathway. *Cell*, **70**, 313.

74. Schindler, C. and Darnell, J. E. Jr. (1995) Transcriptional responses to polypeptide ligands: the JAK–STAT pathway. *Annu. Rev. Biochem.*, **64**, 621.

75. Yan, R., Small, S., Desplan, C., Dearolf, C. R., and Darnell, J. E. Jr. (1996) Identification of a Stat gene that functions in *Drosophila* development. *Cell*, **84**, 421.

76. Copeland, N. G., Gilbert, D. J., Schindler, C., *et al.* (1995) Distribution of the mammalian Stat gene family in mouse chromosomes. *Genomics*, **29**, 225.

77. Muller, M., Laxton, C., Briscoe, J., Schindler, C., Improta, T., Darnell, J. E. Jr., *et al.* (1993) Complementation of a mutant cell line: central role of the 91 kDa polypeptide of ISGF3 in the interferon-alpha and -gamma signal transduction pathways. *EMBO J.*, **12**, 4221.

78. Li, X., Leung, S., Kerr, I. M., and Stark, G. R. (1997) Functional subdomains of STAT2 required for preassociation with the alpha interferon receptor and for signaling. *Mol. Cell. Biol.*, **17**, 2048.

79. Fujitani, Y., Hibi, M., Fukada, T., *et al.* (1997) An alternative pathway for STAT activation that is mediated by the direct interaction between JAK and STAT. *Oncogene*, **14**, 751.

80. Stahl, N., Farruggella, T. J., Boulton, T. G., Zhong, Z., Darnell, J. E. Jr., and Yancopoulos, G. D. (1995) Choice of STATs and other substrates specified by modular tyrosine-based motifs in cytokine receptors. *Science*, **267**, 1349.

81. Heim, M. H., Kerr, I. M., Stark, G. R., and Darnell, J. E. Jr. (1995) Contribution of STAT SH2 groups to specific interferon signaling by the JAK–STAT pathway. *Science*, **267**, 1347.

82. Wen, Z., Zhong, Z., and Darnell, J. E. Jr. (1995) Maximal activation of transcription by Stat1 and Stat3 requires both tyrosine and serine phosphorylation. *Cell*, **82**, 241.

83. David, M., Petricoin, E. R., Benjamin, C., Pine, R., Weber, M. J., and Larner, A. C. (1995) Requirement for MAP kinase (ERK2) activity in interferon alpha- and interferon beta-stimulated gene expression through STAT proteins [see comments]. *Science*, **269**, 1721.

84. Zhang, X., Blenis, J., Li, H. C., Schindler, C., and Chen-Kiang, S. (1995) Requirement of serine phosphorylation for formation of STAT-promoter complexes. *Science*, **267**, 1990.

85. Pfeffer, L. M., Mullersman, J. E., Pfeffer, S. R., Murti, A., Shi, W., and Yang, C. H. (1997) STAT3 as an adapter to couple phosphatidylinositol 3-kinase to the IFNAR1 chain of the type I interferon receptor. *Science*, **276**, 1418.

86. Meraz, M. A., White, J. M., Sheehan, K. C., *et al.* (1996) Targeted disruption of the Stat1 gene in mice reveals unexpected physiologic specificity in the JAK–STAT signaling pathway. *Cell*, **84**, 431.

87. Akira, S. and Kishimoto, T. (1997) NF-IL6 and NF-kappa B in cytokine gene regulation. *Adv. Immunol.*, **65**, 1.

88. Maniatis, T. (1997) Catalysis by a multiprotein IkB kinase complex. *Science*, **278**, 818.

89. Manning, A. and Rao, A. (1997) IKK-1 and IKK-2: cytokine-activated IkB kinases essential for NFκB activation. *Science*, **278**, 860.

90. Woronicz, J. D., Gao, X., Zhaodan, C., Rothe, M., and Goeddel, D. V. (1997) IkB kinase-beta: NFκB activation and complex formation with IkB kinase-alpha and NIK. *Science*, **278**, 866.

91. Zamorano, J., Wang, H. Y., Wang, L. M., Pierce, J. H., and Keegan, A. D. (1996) IL-4 protects cells from apoptosis via the insulin receptor substrate pathway and a second independent signaling pathway. *J. Immunol.*, **157**, 4926.

92. Nagata, S. (1997) Apoptosis by death factor. *Cell*, **88**, 355.

93. Beg, A. A. and Baltimore, D. (1996) An essential role for NF-kappaB in preventing TNF-alpha-induced cell death [see comments]. *Science*, **274**, 782.

94. Van Antwerp, D. J., Martin, S. J., Kafri, T., Green, D. R., and Verma, I. M. (1996) Suppression of TNF-alpha-induced apoptosis by NF-kappaB [see comments]. *Science*, **274**, 787.

95. Wang, C. Y., Mayo, M. W., and Baldwin, A. S. Jr. (1996) TNF- and cancer therapy-induced apoptosis: potentiation by inhibition of NF-kappaB [see comments]. *Science*, **274**, 784.

96. Endo, T. A., Masuhara, M., Yokouchi, M., *et al.* (1997) A new protein containing an SH2 domain that inhibits JAK kinases. *Nature*, **387**, 921.

97. Naka, T., Narazaki, M., Hirata, M., *et al.* (1997) Structure and function of a new STAT-induced STAT inhibitor. *Nature*, **387**, 924.

98. Starr, R., Willson, T. A., Viney, E. M., *et al.* (1997) A family of cytokine-inducible inhibitors of signalling. *Nature*, **387**, 917.

99. Brondello, J. M., Brunet, A., Pouyssegur, J., and McKenzie, F. R. (1997) The dual specificity mitogen-activated protein kinase phosphatase-1 and -2 are induced by the p42/p44MAPK cascade. *J. Biol. Chem.*, **272**, 1368.

100. Boussiotis, V. A., Freeman, G. J., Berezovskaya, A., Barber, D. L., and Nadler, L. M. (1997) Maintenance of human T cell anergy: blocking of IL-2 gene transcription by activated Rap1. *Science*, **278**, 124.

101. Barber, D. L., Mason, J. M., Fukazawa, T., Reedquist, K. A., Druker, B. J., Band, H., *et al.* (1997) Erythropoietin and interleukin-3 activate tyrosine phosphorylation of CBL and association with CRK adaptor proteins. *Blood*, **89**, 3166.

102. Carpino, N., Wisniewski, D., Strife, A., Marshak, D., Kobayashi, R., Stillman, B., *et al.* (1997) p62(dok): a constitutively tyrosine-phosphorylated, GAP-associated protein in chronic myelogenous leukemia progenitor cells. *Cell*, **88**, 197.

103. Yamanashi, Y. and Baltimore, D. (1997) Identification of the Abl- and rasGAP-associated 62 kDa protein as a docking protein, Dok. *Cell*, **88**, 205.

104. Scharenberg, A. M. and Kinet, J. P. (1996) The emerging field of receptor-mediated inhibitory signaling: SHP or SHIP? *Cell*, **87**, 961.

105. Shultz, L. D., Schweitzer, P. A., Rajan, T. V., Yi, T., Ihle, J. N., Matthews, R. J., *et al.* (1993) Mutations at the murine motheaten locus are within the hematopoietic cell protein–tyrosine phosphatase (Hcph) gene. *Cell*, **73**, 1445.

106. Damen, J. E., Liu, L., Rosten, P., Humphries, R. K., Jefferson, A. B., Majerus, P. W., *et al.* (1996) The 145-kDa protein induced to associate with Shc by multiple cytokines is an inositol tetraphosphate and phosphatidylinositol 3,4,5-triphosphate 5-phosphatase. *Proc. Natl. Acad. Sci. USA*, **93**, 1689.

107. Ware, M. D., Rosten, P., Damen, J. E., Liu, L., Humphries, R. K., and Krystal, G. (1996) Cloning and characterization of human SHIP, the 145-kD inositol 5-phosphatase that associates with SHC after cytokine stimulation. *Blood*, **88**, 2833.

3 | Cytokine networks

FIONULA M. BRENNAN AND MARC FELDMANN

1. Introduction

Cytokines are local messenger molecules that transmit information of importance between cells and have a major impact on growth regulation, cell division, differentiation, inflammation, and immunity. In this physiological and pathophysiological context, does it make sense to consider cytokines as a network? We think that this concept *is* useful to help define pathophysiological mechanisms, as we grapple to unravel the complexities of what is happening in normal responses and disease. It is possible to use this concept to point out critical 'rate limiting' steps that may be useful targets for therapeutic intervention.

Immune and inflammatory responses can occur anywhere in the body in response to a wide variety of stimuli. They depend on the rapid and efficient recruitment of a variety of different blood cells, with the exact type and relative proportion dependent upon the type of insult, and the chemotactic and adhesion response generated. In this rapidly changing environment, cytokines are produced that have both agonist and antagonist action. Therefore, the outcome of the immune/inflammatory response depends on the net effect of these mediators. If the pro-inflammatory milieu exceeds that of the anti-inflammatory mediators, the net effect will result in prolonged inflammation.

In this review, we will discuss the evidence that cytokine activities should be considered as a network, rather than individually, and consider how this approach has been helpful in understanding the pathogenesis of disease.

2. Cloning cytokines and evaluating their functions

The impact of molecular biology on the cytokine field has been revolutionary. In the past it has been a tremendous struggle to purify just a few micrograms of cytokine protein from many hundred of millions of cells in order to perform biochemical and functional analyses. Now the cloning of cytokine cDNAs has led to an abundant supply of pure cytokines and reagents (cDNA probes, antibodies, etc.) to study function. However, cytokines are not released by activated cells in isolation and the cells they target are always subject to multiple stimuli. Hence a simple experiment to evaluate the properties of a single cytokine on a clonal cell line *in vitro* may be unrepresentative, especially as cells are often grown in 5–10% fetal calf serum. *In vivo* cytokines act on complex mixtures of cells with the net effect that may not be predictable from *in vitro* experiments. Also, cytokines often act *in vivo* in high concentrations of plasma, which may contain other cytokines and cytokine inhibitors. Serum formed after blood clots also contains the products of platelet granules. Cytokines bioactive in serum include M-colony-stimulating factor (CSF), G-CSF, and interleukin (IL) 6, and levels of these tend to be elevated during an active response to many types of stress. Other cytokines are not usually found in bioactive concentrations, even in severe illness, e.g. tumour necrosis factor (TNF)-α, IL-1, interferon (IFN)-γ. More numerous are cytokine inhibitors in serum, most of which would not be present in fetal calf serum. It is the presence of cytokine inhibitors in plasma that alters the *in vivo* response to cytokines in a critical way. The duration (and extent) of exposure is limited by the inhibitory effect.

The major cytokine inhibitors include a single receptor antagonist (for IL-1), and a large number of soluble cytokine receptors (the extracellular domain) usually derived from the same gene as the transmembrane receptor, but occasionally after alternative splicing of the receptor mRNA. Soluble receptors are physiologically present at high levels, e.g. 20 ng/ml for soluble IL-6 receptor, which far exceeds the concentration of the cytokine. *In vivo*, cytokines are rapidly metabolized, either by kidney or liver, or after internalization by receptors, and this again limits the duration and extent of cytokine exposure.

In conclusion, evaluation of cytokine properties by *in vitro* analysis using cell lines has produced many artefacts and properties that could not occur *in vivo*. Cytokine exposure *in vivo* is more transient and rate limiting, not only because of pulsatile cytokine release from activated cells but also because of the presence of plasma inhibitors and efficient clearance mechanisms.

3. Analysis of cytokine activity *in vivo*

It is thus apparent that physiological and pathophysiological cytokine activities can only be fully appreciated *in vivo*. How can this be studied? A variety of models have been used, including cytokine transgenic and cytokine/cytokine receptor knock-out mice. Cytokine injection and anti-cytokine receptor antibodies, given once or chronically, have also been informative. For humans, analysis of pathological tissue has been useful but very limited information is available about cytokine and anti-cytokine antibody injection.

3.1 Cytokine transgenic mice

The use of various promoters has permitted the local expression of cytokines in various sites with a variety of consequences, some predicted, others not. For example, production of IL-6 under the control of the mouse immunoglobulin promoter led to the proliferation of B lymphocytes and hypergammaglobulinaemia (1). The local production of IFN-γ in the pancreas leads to the recruitment of inflammatory cells, and the development of autoimmune diseases, as previously predicted (2, 3). Thus autoimmune diabetes develops if IFN-γ production is driven in the islet β cells of the pancreas by the insulin promoter (3). The local overproduction of most other cytokines (e.g. TNF-α, IL-2) in the islet β cells is in itself insufficient to generate autoimmunity with autoreactive T cells, etc. (4, 5). It is important to stress that although there is local overproduction of a single cytokine, initially the results are very difficult to interpret, as the recruited cells are activated and will themselves produce or induce multiple cytokines. Thus even a single stimulus generates a cytokine network *in vivo*.

A well-documented example of local cytokine expression are the TNF-α transgenics produced by Kollias and colleagues; this topic is discussed in detail in Section 4.2. The original experiments replaced the 3′ untranslated region of the human TNF-α gene with the globin 3′ untranslated region, and these mice were found to chiefly develop an erosive arthritis, with cachexia and skin disease in some strains (6). The effects of local TNF-α were not direct, but involved a network, with a massive infiltrate of leukocytes, mostly neutrophils. Direct evidence that the pathology depends on a cytokine network has come from therapeutic experiments in these mice. Blocking IL-1 action with an antibody to the IL-1 receptor was as effective in preventing arthritis as a monoclonal antibody to human TNF-α (7). This result emphasizes that some cytokines, e.g. TNF-α, induce the production of other cytokines, e.g. IL-1. Furthermore, it highlights the concept of synergy. These synergies include TNF-α and IL-1, but also IFN-γ and TNF-α. The mechanism of some of these synergies is understood, for example IFN-γ up-regulates TNF receptor expression (8), others are not.

3.2 Cytokine knock-out mice

The *in vivo* function of many cytokines have been clarified by the production and study of cytokine knock-out mice. Many of the results were not predicted by the *in vitro* analysis of function. An interesting example is the IL-2 knock-out mouse (9). These mice develop normally for the first few weeks, then, depending on their genetic background, develop autoimmune manifestations, with haemolytic anaemia and colitis being the most prominent manifestations (10). This pathology was unexpected because it was thought that such mice would have abnormal T cell development. Rather, these results suggest that IL-2 has a major role in the generation of immunoregulatory T cells, although the pathology in IL-2 knock-out animals is not well understood.

Interleukin-10 is a cytokine with many activities (discussed in Section 5.1). It activates B lymphocytes but has powerful inhibitory actions on T cell and macrophage activation, reducing the production of pro-inflammatory cytokines such as TNF-α and IL-1 (see also Chapter 6). Thus, it is not surprising that IL-10 knock-out mice have a variety of problems. The most dramatic is colitis, which has a resemblance to Crohn's disease, and is preventable by treatment with IL-10 (11). This colitis is an exaggerated response to gut micro-organisms, as germ-free IL-10 knock-out mice do not get colitis. It is not certain which pro-inflammatory cytokines are up-regulated in the absence of IL-10, but TNF-α and IL-1 are obvious candidates (reviewed in ref. 12).

The transforming growth factor (TGF)-β family is comprised of three closely related cytokines (TGF-β1, -β2, and -β3) that interact with the same receptor complex (discussed in Section 5.3). TGF-β1 is the most widely expressed, but β2 is also abundant. TGF-β1 knock-out mice were found to develop widespread inflammatory diseases from an early age, and died aged three to four weeks (13).

These results were not predicted because TGF-β2 exhibits similar activity to TGF-β1 *in vitro*, yet clearly does not compensate in the TGF-β1 knock-out mice. Such observations highlight the limitations of fully understanding cytokine function from *in vitro* studies.

Whilst the generation and subsequent analysis of transgenic and knock-out cytokine mice has been informative, these studies have also encountered problems. Not least is the absolute involvement of certain cytokines or their receptors in embryogenesis, leading to 'lethal' knock-outs. Second is the expected problem with respect to susceptibility to infection. However, with rapid advances in technology, 'conditional' knock-outs and transgenics will soon be available that will allow the time-dependent expression or deletion of a given cytokine gene, in addition to tissue-specific expression of the molecule in question. Therefore, the role of a cytokine in the context of a controlled situation could be evaluated much more critically than at present. A variant on knock-out technology are the so-called 'knock-in' mice. In this case the endogenous gene is replaced with a mutated form of the original gene. An example of this is the δ,1–12 TNF gene, which lacks the cleavage site for the TNF convertase enzyme, resulting in membrane-anchored TNF but no secreted cytokine. Such 'knock-in' animals take the complexities of cytokine interactions one step further by examining the relative contribution of membrane-bound versus soluble cytokine function.

4. Regulation of cytokines in rheumatoid arthritis

A broader understanding of cytokine interactions and networks has been derived from studies of cytokine expression and regulation in disease states. Our particular interest has been the role of cytokines in contributing to the chronic inflammatory disease rheumatoid arthritis (RA). RA is a common human autoimmune disease with a prevalence of about 1%, and whilst there has been some progress in defining its aetiology and pathogenesis, these are still incompletely understood (14). RA is characterized by chronic inflammation of the synovial joints and infiltration by blood-derived cells, chiefly memory T cells, macrophages, and plasma cells, all of which show signs of activation (15, 16). This leads in most cases to progressive destruction of cartilage and bone, which occurs after invasion of these tissues by the cellular synovial tissue, and is believed to be mediated by cytokine induction of destructive enzymes, such as matrix metalloproteinases. A wide spectrum of pro-inflammatory and anti-inflammatory cytokines are found in RA synovial tissue, at

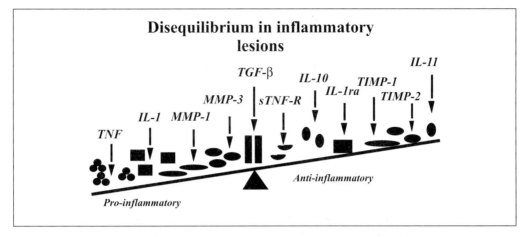

Fig. 1 Cytokine disequilibrium in rheumatoid arthritis. Cytokine profile at sites of inflammation such as rheumatoid synovial tissue is characterized by increased production of both pro- and anti-inflammatory molecules. Modified and reproduced with permission from Cell Press (Feldmann, M., Brennan, F. M., and Maini, R. N. (1996) Rheumatoid arthritis. *Cell*, **85**, 307).

both mRNA and protein level (reviewed in ref. 17). Furthermore, if this synovial tissue is enzymatically dissociated and the cells cultured *in vitro* without extrinsic stimulation, they spontaneously produce many cytokines for a prolonged period of time, indicating that the signals driving cytokine production are present (reviewed in ref. 17). Whilst there is increased production of many pro-inflammatory cytokines, including TNF-α, IL-1, IL-6, and IL-8, this is offset to some degree by increased, albeit insufficient, production of cytokine inhibitors such as IL-10, soluble TNF receptors, and the IL-1 receptor antagonist. Thus, at sites of inflammation, the cytokine profile is characterized by increased synthesis of pro- *and* anti-inflammatory cytokines. The overall effect of these molecules on the pathogenesis of the disease depends on the balance between them, and since the production of the anti-inflammatory molecules is partly a consequence of the synthesis of the pro-inflammatory cytokines, there exists a complex cytokine network. This concept is illustrated in Fig. 1.

4.1 What are the upstream cytokine inducers of the cytokine network in RA?

To investigate the cytokine interactions in this tissue we abrogated the activity of one particular molecule and then determined the 'downstream' effect of that upon synthesis of other cytokines. Initially we focused upon TNF-α, as this cytokine, together with IL-1, were considered to be the most pro-inflammatory and the only ones known to initiate destruction of the underlying cartilage and bone (18–23). Furthermore, it was known that TNF-α was a potent inducer both of itself and of IL-1 (24). By adding anti-TNF-α antibodies to the RA synovial cultures we found that production of bioactive IL-1 was diminished (25). Blockade of TNF-α also reduced the

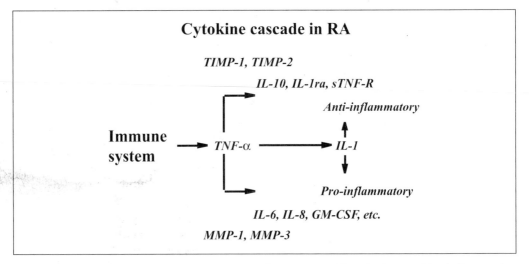

Fig. 2 Cytokine cascade in rheumatoid arthritis. Pro-inflammatory cytokines in rheumatoid arthritic synovium interact in a 'network' or 'cascade'. Modified and reproduced with permission from Cell Press (Feldmann, M., Brennan, F. M., and Maini, R. N. (1996) Rheumatoid arthritis. *Cell*, **85**, 307).

production of the pro-inflammatory cytokines, including granulocyte-macrophage (GM)-CSF, IL-6, and IL-8 (26, 27). In contrast, blocking IL-1 with IL-1 receptor antagonist reduced IL-6 and IL-8 production but not that of TNF-α (27). These results indicate that, at this site of inflammation, the pro-inflammatory cytokines were linked in a 'network' or 'cascade' as illustrated in Fig. 2. It is suggested that blocking TNF-α would also diminish the actions of many other pro-inflammatory mediators, and hence, that TNF-α may be a good therapeutic target.

4.2 Cytokine interactions in animal models of arthritis

These results were to some extent surprising because it had previously been thought that where there was up-regulated production of many different cytokines, blockade of any one cytokine alone would not have much effect (28). However, our initial *in vitro* studies were validated *in vivo* in mouse (DBA/1) collagen-induced arthritis (CIA). Whilst it was known that CIA required both T and B cell activation, the involvement of cytokines including TNF was not known. We and others demonstrated that anti-TNF antibody treatment of DBA/1 mice with *established disease* blocked further joint destruction as measured by clinical evaluation and histological analysis (29, 30). Subsequently, it has been found that virtually all animal models of arthritis are ameliorated by anti-TNF-α, including human TNF-α transgenic arthritis, adjuvant arthritis, streptococcal cell wall arthritis, and antigen-induced arthritis in rabbits (reviewed in ref. 17).

Another informative model of arthritis was developed by G. Kollias and colleagues (31). A human TNF-α transgene was generated in which the AU-rich 3′ UT region (which shortens mRNA half-life) was replaced by the stable β globin 3′ UT, resulting in overexpression of TNF-α. Mice from the original line (Tg 197) were found to chiefly develop an erosive arthritis, which was prevented by treating the animals with neutralizing anti-huTNF-α antibodies. The pathogenesis of the erosive arthritis has been investigated in detail. Of interest is the observation that this transgenic disease is not autoimmune, as TNF-α transgenic mice backcrossed to Rag1 knock-outs (which do not have T or B lymphocytes) also develop an erosive arthritis (32). The onset of arthritis in the transgenic mice will occur if the transgene is either human *or* mouse in origin, and will also develop if the transgene is the non-cleavable form of either human or mouse TNF (δ,1–12 TNF) (33, 34). This suggests that the onset of arthritis in these transgenic animals is not dependent on the secretion of TNF, as membrane TNF will suffice and, further, that the disease involves cytokine cascades. Mouse IL-1 is thought to be particularly important (7).

We have backcrossed the huTNF-α-3′ β globin (Tg 197) mice onto the arthritis-susceptible DBA/1 background and found an acceleration in the age of onset of arthritis with successive generations of interbreeding (35). Using synovial tissue dissected out of the knees of these animals, we have investigated cytokine inter-actions. Bioactive huTNF-α in primary synovial membrane cell cultures was signific-antly higher in the DBA/1 transgenic mice compared with transgenic mice on the original background, and this was accompanied by increases in synovial cell expression of murine IL-1β and IL-6, which supports the *in vivo* findings (7) that a TNF-induced IL-1 network was present. In contrast, murine GM-CSF, IFN-γ, and IL-4 could not be detected, and whilst the anti-inflammatory cytokine IL-10 could be detected, levels were not modulated by expression of the transgene. Of particular relevance was the observation that fibroblasts derived from synovial tissue spon-taneously released the human TNF-α transgene and, using immunohistochemical techniques, this cytokine was localized to fibroblast-like cells and chondrocytes *in vivo*. This finding is of importance because fibroblasts do not normally produce TNF protein due to the presence of a *trans* repressor that binds 3′UTR (36). Secondly, unpublished observations indicate that the absence of 3′UTR in the TNF gene reduces the capacity of IL-10 to regulate its expression (Kollias, personal com-munication). Taken together these studies indicate that disregulated expression of a single cytokine gene can lead to pathology, and that this is partly because the full spectrum of cytokine network interactions is not functional, particularly those involving regulation by IL-10.

4.3 What regulates the cytokine network?

From the studies described above, our current view of the cytokine network in rheumatoid synovium, and probably in other chronic inflammatory sites, involves two key concepts: first, that TNF-α is at the 'apex' of a pro-inflammatory cytokine cascade, as illustrated in Fig. 2, and secondly, that there is up-regulation of multiple anti-inflammatory mediators. Hence the rheumatoid synovium can be envisaged as an equilibrium, just tilted towards the pro-inflammatory side (Fig. 1).

The cytokine inhibitors can be divided into two categories: those cytokines that have immunoregulatory activity such as IL-10, and specific cytokine inhibitors such as IL-1ra or soluble TNF-Rs that bind receptor or ligand and thus block activity. It is particularly important that these cytokines and inhibitors are also connected with the cytokine network, and indeed are up-regulated as a consequence of inflammation.

5. Immunoregulatory cytokines

5.1 Interleukin 10

IL-10 was initially described as a murine Th2 cell product that inhibited cytokine synthesis of Th1 cells (37). It is now known to be produced by a range of cell types (CD4$^+$ Th0 and Th1 cells, CD8$^+$ T cells, B cells, keratinocytes, various tumour cell lines and, of particular importance, monocytes/macrophages) (38, 39) (reviewed in ref. 37). IL-10 inhibits the synthesis of pro-inflammatory cytokines from monocytes including IL-1, IL-6, IL-12, IFN-α, TNF-α, GM-CSF, and G-CSF, but also chemokines such as IL-8 and MIP-α (40–43). In addition to inhibiting TNF-α and IL-1 biosynthesis, IL-10 also induces the production of their natural inhibitors. Thus, IL-10 induces IL-1ra mRNA expression and can synergize with LPS to induce IL-1ra protein release in monocytes. Work at the Kennedy Institute demonstrated that IL-10 induces the release of both soluble TNF receptors, p55 and p75, in human monocytes and in RA synovial membrane cell cultures (44). These results collectively explain how IL-10 exerts strong anti-inflammatory activities and, thus, has been termed a 'macrophage deactivating factor' (45) (Fig. 3). In RA synovial membrane cultures that spontaneously produce IL-10 (46, 47) we found that the endogenous IL-10 produced is functional, because inhibition of its activity with a neutralizing monoclonal antibody enhanced TNF-α and IL-1 production (46). Conversely addition of recombinant IL-10 to these cultures inhibited TNF-α and IL-1 production by approximately 50%. In a similar study, using synovial tissue organ cultures (48), exogenous IL-10 also inhibited IL-1β. However, IL-4 was a more potent inhibitor of IL-1β synthesis, and also induced its native inhibitor the IL-1 receptor antagonist (IL-1ra).

Thus, the anti-inflammatory cytokine IL-10 is of central importance in cytokine networks as it acts to negatively regulate the synthesis of pro-inflammatory cytokines. Clearly, if the pro-inflammatory cytokine cascade was not constrained, inflammation and tissue destruction would rapidly ensue. Indeed, the presence of homeostatic regulatory mechanisms, including IL-10, undoubtedly contributes to the 'chronic' as opposed to 'acute' nature of RA. The importance of IL-10 in controlling homeostasis is illustrated in mice in which the IL-10 gene has been deleted (reviewed in ref. 12). IL-10$^{-/-}$ mice, derived and maintained in specific pathogen-free conditions, develop a CD4$^+$-mediated chronic enterocolitis. These CD4 cells develop a highly polarized Th1-like population, which is prevented with anti-IL-12 antibodies.

Whilst IL-10 is an *essential* component of the cytokine network, the mechanisms involved in regulation of monocyte IL-10 production are not well understood at either the cellular or biochemical level. Bacterial endotoxin is the best characterized and most consistent signal for IL-10 production. A number of studies suggest that

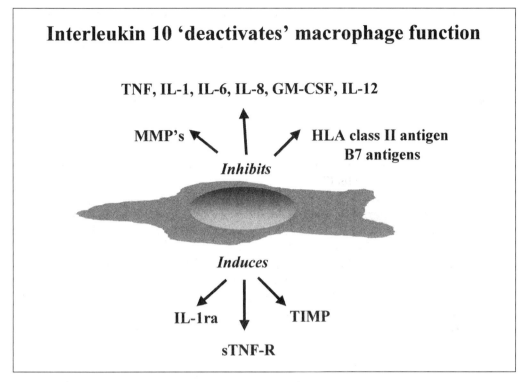

Fig. 3 Interleukin 10 is a macrophage deactivating factor. IL-10 is a potent immunoregulatory molecule deactivating macrophage function that acts principally by inhibiting cytokine synthesis and increasing the production of cytokine inhibitors.

both soluble factors and direct ligation of monocyte cell surface ligands are involved in stimulation of monocyte IL-10 production (49). We and others have shown that TNF-α and IL-1 play a role in the induction of IL-10 by stimulated monocytes (50–52), which suggests the existence of an autoregulatory loop between these pro- and anti-inflammatory cytokines. Indeed, in the RA synovial cultures, we have demonstrated that blockade of TNF-α together with IL-1 reduces the levels of IL-10 (46). However, it is also apparent that whilst these pro-inflammatory cytokines are involved in IL-10 induction, TNF-α-independent pathways also exist and are clearly of importance (49, 52).

5.2 Interleukin 11

There is increasing evidence that IL-11 is another cytokine with anti-inflammatory activity (53, 54). We and others were able to detect elevated levels of IL-11 in synovial membrane mononuclear cell culture supernatants and in synovial fluid of RA patients (55, 56). In our studies on RA synovial membrane cell cultures, exogenous recombinant human IL-11 (rhIL-11) decreased TNF-α levels but only in the presence of an agonistic soluble IL-11 receptor (sIL-11R) (55). In contrast, blocking the biological activity of endogenously produced IL-11 increased TNF-α production about twofold compared with untreated controls, which indicates that IL-11 is active in RA synovial membranes and suppresses TNF-α production. When endogenously produced IL-10 and IL-11 were both blocked in the culture system, TNF-α production increased more than 20-fold, indicating that both cytokines are acting together in synergy as strong inhibitors of the pro-inflammatory cascade *in vivo* (55). Given these effects, the effect of IL-10 on IL-11 synthesis and vice versa was evaluated in RA synovial membrane cultures. Anti-IL-10 antibody reduced IL-11 levels more than twofold, whereas anti-IL-11 antibody increased IL-10 levels twofold, which suggests that IL-10 stimulates the production of IL-11 and that IL-11 down-regulates IL-10 in RA joint cultures. However, experiments on human monocytes showed that IL-11 down-regulates TNF-α by up-regulating IL-10. This difference between the monocyte and RA synovial membrane cell experiment with respect to the mechanism of IL-11 inhibition illustrates the complexities of cytokine interactions. Thus, elevated TNF-α levels in anti-IL-11 antibody treated RA joint cultures may contribute to the increased IL-10 levels found in RA joint cultures. This suggests that the synergistic effect of blocking both IL-11 and IL-10 on TNF-α levels in RA joint cultures is a consequence of the mutual effect of IL-11 and IL-10 on the production of each other, but also to their independent effects on inhibiting TNF-α synthesis, by mechanisms that are as yet not fully defined. The cytokine interactions between the pro-inflammatory cytokines TNF/IL-1 with IL-10 and IL-11 are illustrated in Fig. 4.

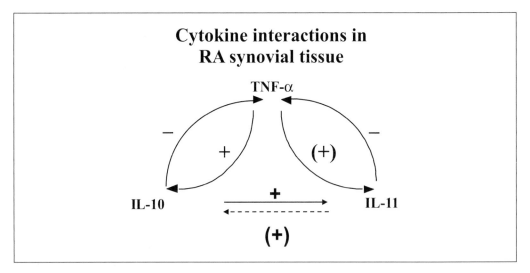

Fig. 4 Theoretical interactions between pro-inflammatory TNF and anti-inflammatory IL-10 and IL-11 in RA synovial tissue. Interactions between these mediators were deduced from experiments performed on RA synovial membrane cultures. Stimulatory effects are indicated as '+', and inhibitory effects as '–'. Dashed lines indicate an indirect effect. Modified and reproduced with permission from *Arthritis and Rheumatism* (Hermann. J. A., Hall, M. A., Maini, R. N., Feldmann, M., and Brennan, F. M. (1998) *Arthritis and Rheumatism*, **41**, 1388).

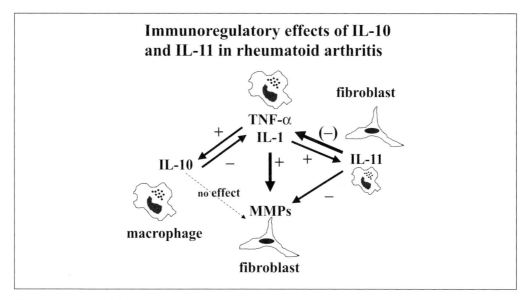

Fig. 5 IL-10 and IL-11 interact to regulate pro-inflammatory effects. They cause both overlapping and independent effects of macrophages and fibroblasts in RA synovial tissue to 'down-regulate' cytokines and enzymes with pro-inflammatory effects, and increase the production of inhibitors.

Cytokine interactions and networks are further illustrated with respect to cytokine receptor distribution. IL-10 and IL-11 both can protect mesenchymal tissue in RA, by inhibiting the synthesis of matrix metalloproteinases (MMPs) such as collagenase in stimulated human monocytes, and by increasing the synthesis of the natural MMP inhibitor TIMP (tissue inhibitor metalloproteinase) (55, 57). However, in RA synovial tissue, the main source of these MMP enzymes are the fibroblasts that lack IL-10 receptors, and thus IL-11 (but not IL-10), and inhibit MMP synthesis in this tissue. In contrast IL-10 will always inhibit macrophage-derived cytokines, whereas the effect of IL-11 on macrophages is dependent upon sufficient expression of the IL-11 α chain receptor. Thus, together, IL-10 and IL-11 regulate cytokine production and pathological effects by (as yet unknown) synergistic interactions and by targeting different cell types. These interactions are illustrated in Fig. 5.

5.3 TGF-β

Members of the transforming growth factor beta family (TGF-β1, TGF-β2, and TGF-β3) also have potent immunoregulatory properties. It is released from an inactive precursor form and binds to cell surface-specific receptors (58–60) and is found in abundance in both inactive and active RA (61–63). However, whereas TGF-β (1 and 2) is present in abundant amounts in RA synovial fluids and is produced spontaneously by RA synovial membranes, its role in the pathogenesis of RA still remains elusive. For example, TGF-β inhibits immune functions including those of T and B cells (64) and reduces TNF-α and IL-1 production in peripheral blood monocytes, but this effect is not detected in RA synovial membrane mononuclear cells (63, 65). TGF-β is chemotactic for neutrophils and monocytes and promotes angiogenesis, which may occur via up-regulation of vascular endothelial growth factor (VEGF) *in vivo* (61, 66–68) and, thus, may contribute to an inflammatory response.

On the other hand, TGF-β is important in tissue protection and repair; down-regulates the production of MMPs such as collagenase; stimulates the production of tissue inhibitors of metalloproteinases (TIMP), and induces the synthesis of collagen type I (62, 67, 69, 70). Thus local TGF-β may promote the reparative processes in arthritic synovial connective tissue, and promotes scarring and tissue repair by inhibiting cartilage and bone destruction. The *in vivo* effects of TGF-β in animal models are also contradictory. Thus, if injected locally into the joints of normal rats, TGF-β resulted in a rapid leukocyte infiltration with synovial hyperplasia leading to synovitis (71), whereas if injected systemically into rodents 'susceptible' to arthritis it antagonized the development of polyarthritis (72, 73). Furthermore, anti-TGF-β antibody, if injected locally into the joint of rats with arthritis, diminished the ongoing disease (74). These studies indicate the multipotential properties of TGF-β, and the differential effects obtained if it is injected systemically or locally into the joint. The inhibitory and stimulatory effects of TGF-β on cells of the immune system may partly be explained by the different effects of TGF-β on resting and activated cells. Resting cells may be stimulated and activated cells may be inhibited by TGF-β, making TGF-β a regulatory cytokine potentially capable of converting an inflammatory site into one dominated by resolution and repair (75). Systemic application of TGF-β is an option in chronic inflammatory diseases such as RA, but its suppressive activity on non-activated immune cells not involved in the inflammatory process and its importance for the defence against micro-organisms, may lead to a higher incidence of infections in these patients. Its activity to stimulate collagen production and to inhibit degradation of mesenchymal tissue may lead to fibrotic conditions in organs such as lungs and liver. Thus, a single cytokine such as TGF-β may both exhibit stimulatory and inhibitory effects and, as such, its contribution to the cytokine network is complex; therefore, its potential use as a therapeutic agent has yet to be investigated.

5.4 Interleukin 4

In common with TGF-β, IL-4 also displays some immunoregulatory effects such as inhibition of LPS-induced IL-1, TNF-α, PGE2, and 92 kDa gelatinase production in macrophages (76–79). However, in contrast to TGF-β or IL-10, IL-4 has not been found in rheumatoid synovial tissue cultures (80), nor in T cells cloned from RA synovial biopsies (81), although it has been detected in reactive arthritis (82). This and other evidence suggests that CD4[+] Th2-derived cytokines are not abundant in RA joints, and that CD4[+] Th1 cells predominate in this site (83). It is possible that the lack of IL-4 producing CD4[+] Th2 cells contributes to the pathogenesis of RA, and this has led to suggestions that IL-4 may be a useful therapeutic agent (Mitchison, unpublished observation) (84). Miossec and his colleagues have also demonstrated that the addition of recombinant IL-4 to RA synovial tissue organ cultures resulted in the inhibition of pro-inflammatory cytokine production (84). Using dissociated synovial cell cultures (85), the addition of IL-4 had no significant effect on TNF-α levels in RA synovial membrane cell cultures, but did enhance the expression of membrane-bound p55 and p75 TNF receptor in RA synovial joint cells. This indicates that IL-4 may even enhance TNF-α activity in RA joints. These effects aside, it is interesting to note that IL-4 production in RA joints is defective, and there is a report suggesting that the incidence of allergies is lower in RA patients (86).

6. Cytokine inhibitors

This heterogeneous family of molecules consist of the IL-1 receptor antagonist and many soluble cytokine receptors. The best studied in this latter group are the shed forms of both the p55 and p75 TNF-R, which have sufficiently high affinity in the monomeric form to bind TNF-α and inactivate it. These TNF inhibitors were first detected in urine and serum almost concurrently in 1988–1989, by the groups of Dayer (87), Olsson (88), and Wallach (89). These were subsequently characterized as the extracellular domains of the two TNF receptors. Using antibodies to these proteins, ELISA assays were developed and it was observed that soluble TNF-R (sTNF-R) were present in normal serum at levels of 1–4 ng/ml, with p55 at a lower level than p75. In RA serum, levels of both were elevated, and even more so in synovial fluid, where levels three to four times those of serum were reached. Furthermore, sTNF-R levels in plasma correlated with disease activity (90, 91). These results of up-regulated TNF inhibitor in RA are of interest, as they help exclude the possibility that a major contribution to the pathogenesis of RA is the failure to produce inhibitory factors normally. Indeed, we observed that in RA synovial cultures, sTNF-Rs were produced at concentrations capable of neutralizing a significant proportion of the TNF-α generated (92). Thus, local production of cytokine inhibitors is capable of diminishing disease activity, and cytokine activity is partially down-regulated by endogenous inhibitors. Furthermore, it is apparent that sTNF-Rs are induced as a consequence of the inflammatory response because TNF-α itself can induce sTNF-R release *in vitro*, and also *in vivo* (93, 94).

Soluble IL-1 receptor has also been detected in RA tissues, initially in synovial fluid (95). This was first found as an IL-1β binding protein, and was subsequently identified using monoclonal antibody as the Type II IL-1R. It is of interest that this receptor is not involved in signalling, and appears to function not only as a decoy on the cell surface, but also as an inhibitor. In its soluble form it binds pro-IL-1β and thus prevents its processing, but it does not bind IL-1ra (96). In addition to the sIL-1-R, the third member of the IL-1 family, the only known cytokine receptor antagonist, IL-1ra (reviewed in ref. 97), is found in abundance in RA synovium (98–100). However, the ratio of IL-1ra to IL-1 in these cultures ranged from 1.2 to 3.6, which is well below the 100-fold excess of IL-1ra needed to neutralize IL-1 bioactivity. Therefore, it is not surprising that bioactive IL-1 is found in the majority of RA culture supernatants (25). Normal joint tissues express very little IL-1ra, so IL-1ra production is up-regulated in the disease process, but not sufficiently to neutralize IL-1. It is of interest that a study of the ratio of IL-1ra:IL-1 in the synovial fluid of patients with Lyme arthritis indicated that the patients with the most favourable outcome had the highest ratio of IL-1ra to IL-1 (101).

7. What maintains the cytokine network in rheumatoid arthritis?

In the preceding sections we have documented how pro- and anti-inflammatory cytokines interact, using studies from RA to illustrate the point. From these and other studies reviewed extensively (17) we now have a clearer understanding of the cytokine players and of the homeostatic mechanisms present. However, if TNF-α is at the top of the cytokine cascade it clearly is important to address the mechanism(s) involved in TNF-α induction, as this would indicate how the cytokine network or cascade is maintained. Furthermore, whilst we have demonstrated that TNF-α is an important signal for the induction of other pro-inflammatory cytokines in RA synovium (17), the mechanism that leads to TNF-α production remains unclear. Based on some earlier unpublished findings, which indicated that synovial monocyte/ macrophage TNF-α production was reduced if the adherent cells were cultured without the non-adherent population (Brennan, unpublished), we focused our attention on the role of T cells, and in particular cognate T cell:macrophage interactions in inducing TNF-α and IL-10 production. Based on a system first described by Jean-Michael Dayer (102) we stimulated peripheral blood lymphocytes with agents to

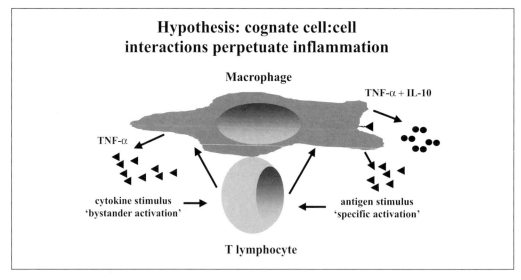

Fig. 6 Cognate interaction of activated T cells with macrophages results in cytokine production. Fixed-activated T cells deliver cell surface signals to macrophages resulting in macrophage-derived cytokine production. 'Bystander' T cell activation leads to TNF synthesis (but not IL-10); 'antigen-dependent' T cell activation results in TNF *and* IL-10 synthesis.

mimic antigen activation. T cells were fixed so that they neither released soluble factors, or could respond themselves, and were presented to monocytes in culture. Using this co-culture system we observed that monocyte-derived TNF-α and IL-10 production occurred when the T cells were stimulated via the T cell receptor (49). However, if the T cells were activated in an antigen-independent manner using a cocktail of T cell stimulatory cytokines that included IL-15, TNF-α, but *not* IL-10, production occurred (103). This difference could be of importance for understanding the balance between pro- and anti-inflammatory cytokine production in RA synovial tissue in which there are abundant T cells and macrophages. Thus, TNF-α (but not IL-10) synthesis follows the interaction of T cells activated by macrophage-derived cytokines with macrophages themselves. This is illustrated in Fig. 6. A role for IL-15 in RA has also been proposed by Liew and colleagues who detected IL-15 in RA synovium and in synovial fluid (104, 105).

8. Demonstration of cytokine network interactions *in vivo*

8.1 Immunotherapy of RA

The down-regulation of pro-inflammatory cytokines by anti-TNF-α antibody in RA joint cultures *in vivo*, and the benefit of anti-TNF-α in collagen-induced arthritis, provided the rationale for clinical trials of anti-TNF-α antibody, in long-standing RA patients with active disease. The results of multiple clinical trials of anti-TNF-α antibody performed with cA2, a chimeric anti-TNF-α antibody (mouse Fv, human IgG1), have revealed a marked benefit in all assessable aspects of the disease. There was very rapid improvement within days, and the benefit lasted from 8–26 weeks in the first trial (106). After these patients relapsed, retreatment with cA2 ameliorated disease on each occasion, which indicates that the pro-inflammatory cascade remains TNF-α dependent. The clinical efficacy of anti-TNF-α has been verified in randomized, double-blind, placebo-controlled trials (107). These results have triggered clinical trials with other anti-TNF-α reagents; a different anti-TNF-α antibody of IgG4 isotype (108) and TNF receptor IgG fusion proteins (109). All of these have been successful. In terms of insights into mechanism of disease, the results of these clinical trials indicate that while there may be diversity of immune responses in heterogeneity in different populations, as suggested by some human lymphocyte antigens (HLA)-DR, there is little heterogeneity in terms of cytokine mechanisms. These results emphasize that cytokines are good therapeutic targets in inflammatory/autoimmune diseases.

The clinical trials have provided an opportunity to learn more about the disease process and the mechanism by which anti-TNF-α antibody works. First, it was shown that the elevated serum IL-6 levels fall rapidly, normalizing in a day or so. This verifies the operation of a TNF-α-dependent cytokine cascade *in vivo* (106). Histological analysis of joints after anti-TNF-α therapy revealed diminished cellularity of joints as well as diminished expression of endothelial adhesion molecules

(110). Soluble E-selectin and ICAM-1 in serum were down-regulated and counts of long-lived blood cells such as lymphocytes were rapidly augmented after anti-TNF-α treatment (111). These results suggest that anti-TNF-α antibody may exert a pro-longed benefit because it diminished cell recruitment into joints (reviewed in ref. 112).

However, anti-TNF-α therapy of rheumatoid arthritis is not a cure, and other therapeutic approaches are needed. In collagen-induced arthritis in DBA/1 mice, there is marked synergy between anti-TNF-α antibody and therapy targeted at T cells, for example with anti-CD4 antibody (113). Synergistic approaches, targeting two key aspects of the pathogenesis, may be the best approach towards achieving a cure.

9. Concluding remarks

Despite the abundance of pro-inflammatory cytokines present in an inflammatory site, these are not independently regulated, and there is evidence for a 'cytokine network' *in vitro* and *in vivo*. Judicious choice of where to block the network can lead to clinical benefit; an example of this in rheumatoid arthritis is described. It seems likely that in other diseases other cytokines will be therapeutic targets.

References

1. Suematsu, S., Matsusaka, T., Matsuda, T., Ohno, S., Miyazaki, J., Yamamura, K., *et al.* (1992) Generation of plasmacytomas with the chromosomal translocation t(12;15) in interleukin 6 transgenic mice. *Proc. Natl. Acad. Sci. U.S.A.*, **89**, 232.
2. Bottazzo, G. F., Pujol-Borrell, R., Hanafusa, T., and Feldmann, M. (1983) Hypothesis: Role of aberrant HLA-DR expression and antigen presentation in the induction of endocrine autoimmunity. *Lancet*, **ii**, 1115.
3. Sarvetnick, N., Shizuru, J., Liggitt, D., Martin, L., McIntyre, B., Gregory, A., *et al.* (1990) Loss of pancreatic islet tolerance induced by β-cell expression of interferon-γ. *Nature*, **346**, 844.
4. Allison, J., Malcolm, L., Chosich, N., and Miller, J. F. A. P. (1992) Inflammation but not autoimmunity occurs in transgenic mice expressing constitutive levels of IL-2 in islet β cells. *Eur. J. Immunol.*, **22**, 1115.
5. Wogensen, L., Huang, X., and Sarvetnick, N. (1993) Leukocyte extravasation into the pancreatic tissue in transgenic mice expressing interleukin 10 in the islets of Langerhans. *J. Exp. Med.*, **178**, 175.
6. Probert, L., Keffer, J., Corbella, P., Cazlaris, H., Patsavoudi, E., Stephens, S., *et al.* (1993) Wasting, ischemia, and lymphoid abnormalities in mice expressing T cell-targeted human tumor necrosis factor transgenes. *J. Immunol.*, **151**, 1894.
7. Probert, L., Plows, D., Kontogeorgos, G., and Kollias, G. (1995) The type I interleukin-1 receptor acts in series with tumor necrosis factor (TNF) to induce arthritis in TNF-transgenic mice. *Eur. J. Immunol.*, **25**, 1794.
8. Aggarwal, B. B., Eessalu, T. E., and Hass, P. E. (1985) Characterization of receptors for human tumour necrosis factor and their regulation by gamma-interferon. *Nature*, **318**, 665.
9. Schorle, H., Holtschke, T., Hunig, T., Schimpl, A., and Horak, I. (1991) Development and

function of T cells in mice rendered interleukin-2 deficient by gene targeting. *Nature*, **352,** 621.

10. Horak, I., Lohler, J., Ma, A., and Smith, K. A. (1995) Interleukin-2 deficient mice: A new model to study autoimmunity and self-tolerance. *Immunol. Rev.*, **148,** 35.

11. Kuhn, R., Lohler, J., Rennick, D., Rajewsky, K., and Muller, W. (1993) Interleukin-10-deficient mice develop chronic enterocolitis. *Cell*, **75,** 263.

12. Rennick, D. M., Fort, M. M., and Davidson, N. J. (1997) Studies with IL-10-/- mice: an overview. *J. Leukoc. Biol.*, **61,** 389.

13. Shull, M. M., Ormsby, I., Kier, A. B., Pawlowski, S., Diebold, R. J., Yin, M., *et al.* (1992) Targeted disruption of the mouse transforming growth factor-beta 1 gene results in multifocal inflammatory disease. *Nature*, **359,** 693.

14. Ziff, M. (1990) Rheumatoid arthritis - Its present and future. *J. Rheumatol.*, **17,** 127.

15. Janossy, G., Panayai, G., Duke, O., Bofill, M., Poulter, L. W., and Goldstein, G. (1981) Rheumatoid arthritis: a disease of T-lymphocyte/macrophage immunoregulation. *Lancet*, **ii,** 839.

16. Cush, J. J., and Lipsky, P. E. (1988) Phenotypic analysis of synovial tissue and peripheral blood lymphocytes isolated from patients with rheumatoid arthritis. *Arthritis Rheum.*, **31,** 1230.

17. Feldmann, M., Brennan, F. M., and Maini, R. N. (1996) Role of cytokines in rheumatoid arthritis. *Annu. Rev. Immunol.*, **14,** 397.

18. Gowen, M., Wood, D. D., Ihrie, E. J., McGuire, M. K. B., and Russell, R. G. (1983) An interleukin-1 like factor stimulates bone resorption *in vitro. Nature*, **306,** 378.

19. Saklatvala, J. (1985) TNFα stimulates resorption and inhibits synthesis of proteoglycan in cartilage. *Nature*, **322,** 547.

20. Saklatvala, J., Sarsfield, S. J., and Townsend, Y. (1985) Purification of two immunologically different leucocyte proteins that cause cartilage resorption lymphocyte activation and fever. *J. Exp. Med.*, **162,** 1208.

21. Dayer, J. M., Beutler, B., and Cerami, A. (1985) Cachectin/tumor necrosis factor stimulates collagenase and prostaglandin E2 production by human synovial cells and dermal fibroblasts. *J. Exp. Med.*, **162,** 2163.

22. Dayer, J.-M., de Rochemonteix, B., Burrus, B., Semczuk, S., and Dinerello, C. A. (1986) Human recombinant interleukin 1 stimulates collagenase and prostaglandin E$_2$ production by human synovial cells and dermal fibroblasts. *J. Exp. Med.*, **77,** 645.

23. Thomas, B. M., Mundy, G. R., and Chambers, T. J. (1987) Tumour necrosis factor α and β induce osteoblastic cells to stimulate osteoclast bone resorption. *J. Immunol.*, **138,** 775.

24. Nawroth, P. P., Bank, I., Handley, D., Cassimeris, J., Chess, L., and Stern, D. (1986) Tumor necrosis factor/cachectin interacts with endothelial cell receptors to induce release of interleukin 1. *J. Exp. Med.*, **163,** 1363.

25. Brennan, F. M., Chantry, D., Jackson, A., Maini, R., and Feldmann, M. (1989) Inhibitory effect of TNF α antibodies on synovial cell interleukin-1 production in rheumatoid arthritis. *Lancet*, **2,** 244.

26. Haworth, C., Brennan, F. M., Chantry, D., Turner, M., Maini, R. N., and Feldmann, M. (1991) Expression of granulocyte-macrophage colony-stimulating factor in rheumatoid arthritis: regulation by tumor necrosis factor-alpha. *Eur. J. Immunol.*, **21,** 2575.

27. Butler, D. M., R.N., M., Feldmann, M., and Brennan, F. M. (1995) Modulation of proinflammatory cytokine release in rheumatoid synovial membrane cell cultures. Comparison of monoclonal anti-TNFα antibody with the IL-1 receptor antagonist. *Eur. Cytokine Netw.*, **6,** 225.

28. Arend, W. P., and Dayer, J. M. (1990) Cytokines and cytokine inhibitors or antagonists in rheumatoid arthritis. *Arthritis Rheum.*, **33,** 305.

29. Williams, R. O., Feldmann, M., and Maini, R. N. (1992) Anti-tumor necrosis factor ameliorates joint disease in murine collagen-induced arthritis. *Proc. Nat. Acad. Sci. USA*, **89,** 9784.

30. Thorbecke, G. J., Shah, R., Leu, C. H., Kuruvilla, A. P., Hardison, A. M., and Palladino, M. A. (1992) Involvement of endogenous tumour necrosis factor α and transforming growth factor β during induction of collagen type II arthritis in mice. *Proc. Natl. Acad. Sci. USA*, **89,** 7375.

31. Keffer, J., Probert, L., Cazlaris, H., Georgopoulos, S., Kaslaris, E., Kioussis, D., *et al.* (1991) Transgenic mice expressing human tumour necrosis factor: a predictive genetic model of arthritis. *EMBO J.*, **10,** 4025.

32. Douni, E., Akassoglou, K., Alexopoulou, L., Georgopoulos, S., Haralambous, S., and Hill, S., (1995) Transgenic and knockout analyses of the role of TNF in immune regulation and disease pathogenesis. *J. Inflamm.*, **47,** 27.

33. Georgopoulos, S., Plows, D., and Kollias, G. (1996) Transmembrane TNF is sufficient to induce localized tissue toxicity and chronic inflammatory arthritis in transgenic mice. *J. Inflamm.*, **46,** 86.

34. Alexopoulou, L., Pasparakis, M., and Kollias, G. (1997) A murine transmembrane tumor necrosis factor (TNF) transgene induces arthritis by cooperative p55/p75 TNF receptor signaling. *Eur. J. Immunol.*, **27,** 2588.

35. Butler, D. M., Malfait, A. M., Mason, L. J., Warden, P. J., Kollias, G., Maini, R. N., *et al.* (1997) DBA/1 mice expressing the human TNF-alpha transgene develop a severe, erosive arthritis: characterization of the cytokine cascade and cellular composition. *J. Immunol.*, **159,** 2867.

36. Kruys, V., Kemmer, K., Shakov, A., Jongeneel, V., and Beutler, B. (1992) Constitutive activity of the tumor necrosis factor promoter is canceled by the 3' untranslated region in nonmacrophage cell lines; a trans-dominant factor overcomes this suppressive effect. *Proc. Natl. Acad. Sci. USA*, **89,** 673.

37. Moore, K. W., O'Garra, A., de Waal Malefyt, R., Vieira, P., and Mosmann, T. R. (1993) Interleukin-10. *Annu. Rev. Immunol.*, **11,** 165.

38. Yssel, H., De Waal Malefyt, R., Roncarolo, M. G., Abrams, J. S., Lahesmaa, R., Spits, H., *et al.* (1992) IL-10 is produced by subsets of human CD4+ T cell clones and peripheral blood T cells. *J. Immunol.*, **149,** 2378.

39. O'Garra, A., Stapleton, G., Dhar, V., Pearce, M., Schumacher, J., Rugo, H., *et al.* (1990) Production of cytokines by mouse B cells: B lymphomas and normal B cells produce interleukin 10. *Int. Immunol.*, **2,** 821.

40. Fiorentino, D. F., Zlotnik, A., Vieira, P., Mosmann, T. R., Howard, M., Moore, K. W., *et al.* (1991) IL-10 acts on the antigen-presenting cell to inhibit cytokine production by Th1 cells. *J. Immunol.*, **146,** 3444.

41. Fiorentino, D. F., Zlotnik, A., Mosmann, T. R., Howard, M., and O'Garra, A. (1991) IL-10 inhibits cytokine production by activated macrophages. *J. Immunol.*, **147,** 3815.

42. De Vaal-Malefyt R, Haanen J, Spits H, Roncarolo M-G, Te Velde A, Figdor C, *et al.* (1991) Interleukin-10 (IL-10) and viral IL-10 strongly reduce antigen-specific human T cell proliferation by diminishing the antigen-presenting capacity of monocytes via downregulation of class II major histocompatibility complex expression. *J. Exp. Med.*, **174,** 915.

43. De Waal Malefyt, R., J. Abrams, B. Bennett, C.G. Figdor, and J.E. de Vries. (1991) Interleukin 10(IL-10) inhibits cytokine synthesis by human monocytes: an autoregulatory role of IL-10 produced by monocytes. *J. Exp. Med.*, **174,** 1209.

44. Joyce, D. A., Gibbons, D., Green, P., Feldmann, M., and Brennan, F. M. (1994) Two inhibitors of pro-inflammatory cytokine release, IL-10 and IL-4, have contrasting effects on release of soluble p75 TNF receptor by cultured monocytes. *Eur. J. Immunol.*, **24,** 2699.

45. Bogdan, C., Paik, J., Vodovotz, Y., and Nathan, C. (1992) Contrasting mechanisms for suppression of macrophage cytokine release by transforming growth factor-beta and interleukin-10. *J. Biol. Chem.*, **267,** 23301.

46. Katsikis, P. D., C.Q. Chu, F.M. Brennan, R.N. Maini, and M. Feldmann. (1994) Immuno-regulatory role of interleukin 10 in rheumatoid arthritis. *J. Exp. Med.*, **179,** 1517.

47. Cush, J., Splawski JB, Thomas R, McFarlin JE, Schulze-Koops H, Davis LS, *et al.* (1995) Elevated interleukin-10 levels in patients with rheumatoid arthritis. *Arthritis Rheum.*, **38,** 96.

48. Chomarat, P., Vannier, E., Dechanet, J., Rissoan, M. C., Banchereau, J., Dinarello, C. A., *et al.* (1995) Balance of IL-1 receptor antagonist/IL-1 beta in rheumatoid synovium and its regulation by IL-4 and IL-10. *J. Immunol.*, **154,** 1432.

49. Parry, S. L., Sebbag, M., Brennan, F. M., and Feldmann, M. (1997) Contact with T cells modulates monocyte IL-10 production by TNFα dependent mechanisms. *J. Immunol.*, **158,** 3673.

50. Wanidworanun, C., and Strober, W. (1993) Predominant role of tumor necrosis factor α in human monocyte IL-10 synthesis. *J. Immunol.*, **151,** 6853.

51. Platzer, C., Meisel, C., Vogt, K., Platzer, M., and Volk, H. D. (1995) Up-regulation of monocytic IL-10 by tumor necrosis factor-alpha and cAMP elevating drugs. *Int. Immunol.*, **7,** 517.

52. Foey, A. D., Parry, S. L., Williams, L. M., Feldmann, M., Foxwell, B. M. J., and Brennan, F. M. (1998) Regulation of monocyte IL-10 production by endogenous IL-1 and TNF: Role of the p38 and p42/44 MAP kinases. *J. Immunol.*, **160,** 920.

53. Trepicchio, W. L., M. Bozza, G. Pedneault, and A.J. Dorner. (1996) Recombinant human IL-11 attenuates the inflammatory response through down-regulation of proinflamma-tory cytokine release and nitric oxide production. *J. Immunol.*, **157,** 3627.

54. Redlich, C. A., X. Gao, S. Rockwell, M. Kelley, and J.A. Elias. (1996) IL-11 enhances survival and decreases TNF production after radiation-induced thoracic injury. *J. Immunol.*, **157,** 1705.

55. Hermann, J., Hall, M., Maini, R., Feldmann, M., and Brennan, F. (1998) Important immunoregulatory role of interleukin-11 in the inflammatory process in rheumatoid arthritis. *Arth. & Rheum.*, **41,** 1388.

56. Okamoto, H., M. Yamamura, Y. Morita, S. Harada, H. Makino, and Z. Ota. (1997) The synovial expression and serum levels of interleukin-6, interleukin-11, leukemia inhibitory factor, and oncostatin M in rheumatoid arthritis. *Arthritis Rheum.*, **40,** 1096.

57. Lacraz, S., L.P. Nicod, R. Chicheportiche, H.G. Welgus, and J.M. Dayer. (1995) IL-10 inhibits metalloproteinase and stimulates TIMP-1 production in human mononuclear phagocytes. *J. Clin. Invest.*, **96,** 2304.

58. Massague, J. (1987) The TGFβ1 family of growth and differentiation factors. *Cell*, **49,** 437.

59. Massague, J. (1992) Receptors for the TGFβ1 family. *Cell*, **69,** 1067.

60. Lawrence, D. (1996) Transforming growth factor-β: a general review. *Eur. Cytokine Netw.*, **7,** 363.

61. Fava, R., Olsen NJ, Postlethwaite AE, Broadley KN, Davidson JM, Nanney LB, *et al.* (1991) Transforming growth factor β1 (TGF-β1) induced neutrophil recruitment to synovial tissues: Implications for TGF-β-driven synovial inflammation and hyperplasia. *J. Exp. Med.*, **173,** 1121.

62. Lafyatis, R., Thomson NL, Remmers ER, Flanders KC, Roche NS, and Kim SJ. (1989) Transforming growth factor-β production by synovial tissues from rheumatoid patients and streptococcal cell wall arthritis rats. Studies on secretion by synovial fibroblast-like cells and immunohistological localisation. *J. Immunol.*, **143,** 1142.

63. Brennan, F. M., Chantry, D., Turner, M., Foxwell, B., Maini, R. N., and Feldmann, M.

(1990) Detection of transforming growth factor-beta in rheumatoid arthritis synovial tissue: lack of effect on spontaneous cytokine production in joint cell cultures. *Clin. Exp. Immunol.*, **81,** 278.

64. Shalaby, M., and Ammannn AJ (1988) Suporession of immune cell function in vitro by recombinant human TGFβ. *Cell Immunol.*, **112,** 343.

65. Chantry, D., Turner M, Abney ER, and Feldmann M. (1989) Modulation of cytokine production by transforming factor β1. *J. Immunol.*, **142,** 4295.

66. Wahl, S., Hunt DA, Wakefield LM, McCartney-Francis N, Wahl LM, Roberts AB, *et al.* (1987) Transforming growth factor type β induces monocyte chemotaxis and growth factor production. *Proc. Natl. Acad. Sci. USA*, **84,** 5788.

67. Roberts, A., Sporn MB, Assoian RK, Smith JM, Roche NS, Wakefield LM, *et al.* (1986) Transforming growth factor β: rapid induction of fibrosis and angiogenesis in vivo and collagen formation in vitro. *Proc. Natl. Acad. Sci. USA*, **83,** 4167.

68. Pertovaara, L., Kaipainen A, Nustonen T, Orpana A, Ferrara N, Saksela O, *et al.* (1994) Vascular endothelial growth factor is induced in response to transforming growth factor-beta in fibroblastic and epithelial cells. *J. Biol. Chem.*, **269,** 6271.

69. Wright, J., Cawston TE, and Hazelman BL. (1991) Transforming growth factor β stimulates the production of the tissue inhibitor of metalloproteinases (TIMP) by human synovial and skin fibroblasts. *Biochem. Biophys. Acta*, **1094,** 207.

70. Khalil, N., Bereznay, O., Sporn, M., and Greenberg, A. H. (1989) Macrophage production of transforming growth factor beta and fibroblast collagen synthesis in chronic pulmonary inflammation. *J. Exp. Med.*, **170,** 727.

71. Allen, J. B., Manthey, C. L., Hand, A., Ohura, K., Ellingsworth, L., and Wahl, S. M. (1990) Transforming growth factor-β induces human T lymphocyte migration in vitro. *J. Exp. Med.*, **171,** 231.

72. Brandes, M. E., Allen, J. B., Ogawa, Y., and Wahl, S. M. (1991) Transforming growth factor β1 suppresses acute and chronic arthritis in experimental animals. *J. Clin. Invest.*, **87,** 1108.

73. Kuruvilla, A. P., Shah, R., Hochwald, G. M., Liggit, H. D., Palladino, M. A., and Thorbecke, G. J. (1991) Protective effect of transforming growth factor β1 on experimental autoimmune diseases in mice. *Proc. Natl. Acad. Sci. USA*, **88,** 2918.

74. Wahl, S., Allen JB, Costa GL, Wong HL, and Dasch JR. (1993) Reversal of acute and chronic synovial inflammation by anti-transforming growth factor β. *J. Exp. Med.*, **177,** 225.

75. McCartney-Francis, N., and Wahl, SM. (1994) TGF-β: a matter of life and death. *J. Leuk. Biol.*, **55,** 401.

76. Hart, P. H., Vitti, G. F., Burgess, D. R., Whitty, G. A., Piccoli, D. S., and Hamilton, J. H. (1989) Potential anti-inflammatory effects of interleukin-4: Suppression of human monocyte tumour necrosis factor α, interleukin-1, and prostaglandin F2. *Proc. Natl. Acad. Sci. USA*, **86,** 3803.

77. Essner, R., Rhoades, K., McBride, W. H., Morton, D. L., and Economou, J. (1989) IL-4 down-regulates IL-1 and TNF gene expression in human monocytes. *J. Immunol.*, **142,** 3857.

78. TeVelde, A. A., Huijbens, K., and Heije, J. E. (1989) Interleukin-4 (IL-4) inhibits secretion of IL-1β, tumour necrosis factor α, and IL-6 by human monocytes. *Blood*, **6,** 1392.

79. Lacraz, S., Nicod, I., Galve-de Rochemonteux, B., Baumberger, C., Dayer, J. M., and Weleus, H. G. (1992) Suppression of metalloproteinase biosynthesis in human alveolar macrophages by interleukin-4. *J. Clin. Invest.*, **90,** 382.

80. Miossec, P., Naviliat M, D'Angeac AD, Sany J, and Banchereau J. (1990) Low levels of interleukin-4 and high levels of transforming growth factor β in rheumatoid synovitis. *Arthritis Rheum.*, **33,** 1180.

81. Cohen, S. B. A., Katsikis, P. D., Chu, C. Q., Thomssen, H., Webb, L. M. C., Maini, R. N., *et*

al. (1995) High level of IL-10 production by the activated T cell population within the rheumatoid synovial membrane. *Arthr. & Rheum.*, **38,** 946.

82. Simon, A. K., Seipelt, E., and Sieper, J. (1994) Divergent T-cell cytokine patterns in inflammatory arthritis. *Proc. Natl. Acad. Sci. USA*, **91,** 8562.

83. Miltenburg, A. J., van Laar, J. M., de Kuiper, R., Daha, M. R., and Breedveld, F. C. (1992) T cells cloned from human rheumatoid synovial membrane functionally represent the Th1 subset. *Scand. J. Immunol.*, **35,** 603.

84. Miossec, P., Briolay, J., Dechanet, J., Wijdenes, J., Martinez-Valdez, H., and Banchereau, J. (1992) The inhibition of the production of proinflammatory cytokines and immuno-globulins by interleukin-4 in an ex vivo model of rheumatoid synovitis. *Arth. & Rheum.*, **35,** 874.

85. Cope, A. P., Gibbons, D. L., Aderka, D., Foxwell, B. M., Wallach, D., Maini, R. N., *et al.* (1993) Differential regulation of tumour necrosis factor receptors (TNF-R) by IL-4; upregulation of P55 and P75 TNF-R on synovial joint mononuclear cells. *Cytokine*, **5,** 205.

86. Lewis-Faning, E. (1950) Report on an enquiry into the aetiological factors associated with rheumatoid arthritis. *Ann. Rheum. Dis.*, **91,** 1.

87. Seckinger, P., Isaaz, S., and Dayer, J. M. (1988) A human inhibitor of tumor necrosis factor α. *J. Exp. Med.*, **167,** 1511.

88. Olsson, I., Lantz, S., Nilsson, E., Peetre, C., Thysell, H., Grubb, A., *et al.* (1989) Isolation and characterization of a tumor necrosis factor binding protein from urine. *Eur. J. Haematol.*, **42,** 270.

89. Engelmann, H., Aderka, D., Rubinstein, M., Rotman, D., and Wallach, D. (1989) A tumor necrosis factor-binding protein purified to homogeneity from human urine protects cells from tumor necrosis factor toxicity. *J. Biol. Chem.*, **264,** 11974.

90. Cope, A. P., Aderka, D., Doherty, M., Engelmann, H., Gibbons, D., Jones, A. C., *et al.* (1992) Increased levels of soluble tumor necrosis factor receptors in the sera and synovial fluid of patients with rheumatic diseases. *Arthritis Rheum.*, **35,** 1160.

91. Roux-Lombard, P., Punzi, L., Hasler, F., Bas, S., Todesco, S., Gallati, H., *et al.* (1993) Soluble tumor necrosis factor receptors in human inflammatory synovial fluids. *Arthritis Rheum.*, **36,** 485.

92. Brennan, F. M., Gibbons, D., Cope, A., Katsikis, P., Maini, R. N., and Feldmann, M. (1995) TNF inhibitors are produced spontaneously by rheumatoid and osteoarthritic synovial joint cell cultures: evidence of feedback control of TNF action. *Scand. J. Immunol.*, **42,** 158.

93. Pennica, D., Kohr, W. J., Fendly, B. M., Shire, S. J., Raab, H. E., Borchardt, P. E., *et al.* (1992) Characterization of a recombinant extracellular domain of the type 1 tumor necrosis factor receptor: evidence for tumor necrosis factor-α induced receptor aggregation. *Biochemistry*, **31,** 1134.

94. Lantz, M., Malik, S., Slevin, M. L., and Olsson, I. (1990) Infusion of tumor necrosis factor (TNF) causes an increase in circulating TNF-binding protein in humans. *Cytokine*, **2,** 402.

95. Symons, J. A., Eastgate, J. A., and Duff, G. W. (1991) Purification and characterization of a novel soluble receptor for interleukin-1. *J. Exp. Med.*, **174,** 1251.

96. Symons, J. A., Young, P. R., and Duff, G. W. (1995) Soluble type II interleukin 1 (IL-1) receptor binds and blocks processing of IL-1 beta percursor and loses affinity for IL-1 receptor antagonist. *Proc. Natl. Acad. Sci. USA*, **92,** 1714.

97. Arend, W. P., Welgus, H. G., Thompson, R. C., and Eisenberg, S. P. (1990) Biological propeties of recombinant human monocyte-derived interleukin 1 receptor antagonist. *J. Clin. Invest.*, **85,** 1694.

98. Roux-Lombard, P., Modoux, C., Vischer, T., Grassi, J., and Dayer, J. M. (1992) Inhibitors of interleukin 1 activity in synovial fluids and in cultured synovial fluid mononuclear cells. *J. Rheumatol.*, **19,** 517.

99. Firestein, G. S., Berger, A. E., Tracey, D. E., Chosay, J. G., Chapman, D. L., Paine, M. M., *et al.* (1992) IL-1 receptor antagonist protein production and gene expression in rheumatoid arthritis and osteoarthritis synovium. *J. Immunol.*, **149**, 1054.

100. Malyak, M., Swaney, R. E., and Arend, W. P. (1993) Levels of synovial fluid interleukin-1 receptor antagonist in rheumatoid arthritis and other arthropathies. *Arthritis Rheum.*, **36**, 781.

101. Miller, L. C., Lynch, E. A., Isa, S., Logan, J. W., Dinarello, C. A., and Steere, A. C. (1993) Balance of synovial fluid IL-1β and IL-1 receptor antagonist and recovery from Lyme arthritis. *Lancet*, **341**, 146.

102. Isler, P., Vey, E., Zhang, J. H., and Dayer, J. M. (1993) Cell surface glycoproteins expressed on activated human T cells induce production of interleukin-1 beta by monocytic cells: a possible role of CD69. *Eur. Cytokine Netw.*, **4**, 15.

103. Sebbag, M., Parry, S. L., Brennan, F. M., and Feldmann, M. (1997) Cytokine stimulation of T lymphocytes regulates their capacity to induce monocyte production of TNFα but not IL-10: possible relevance to pathophysiology of rheumatoid arthritis. *Eur. J. Immunol.*, **27**, 624.

104. McInnes, I. B., al Mughales, J., Field, M., Leung, B. P., Huang, F. P., Dixon, R., *et al.* (1996) The role of interleukin-15 in T-cell migration and activation in rheumatoid arthritis. *Nat. Med.*, **2**, 175.

105. McInnes, I. B., Leung, B. P., Sturrock, R. D., Field, M., and Liew, F. Y. (1997) Interleukin-15 mediates T cell-dependent regulation of tumor necrosis factor-alpha production in rheumatoid arthritis [see comments]. *Nat. Med.*, **3**, 189.

106. Elliott, M. J., Maini, R. N., Feldmann, M., Long-Fox, A., Charles, P., Katsikis, P., *et al.* (1993) Treatment of rheumatoid arthritis with chimeric monoclonal antibodies to TNFα. *Arthritis Rheum.*, **36**, 1681.

107. Elliott, M. J., Maini, R. N., Feldmann, M., Kalden, J. R., Antoni, C., Smolen, J. S., *et al.* (1994) Randomised double blind comparison of a chimaeric monoclonal antibody to tumour necrosis factor α (cA2) versus placebo in rheumatoid arthritis. *Lancet*, **344**, 1105.

108. Rankin, E. C. C., Choy, E. H. S., Kassimos, D., Kingsley, G. H., Sopwith, S. M., Isenberg, D. A., *et al.* (1995) The therapeutic effects of an engineered human anti-tumour necrosis factor alpha antibody (CD571) in rheumatoid arthritis. *Br. J. Rheumatol.*, **34**, 334.

109. Moreland, L. W., Baumgartner, S. W., Schiff, M. H., Tindall, E. A., Fleischmann, R. M., Weaver, A. L., *et al.* (1997) Treatment of rheumatoid arthritis with a recombinant human tumor necrosis factor receptor (p75)-Fc fusion protein. *N. Engl. J. Med.*, **337**, 141.

110. Tak, P. P., Taylor, P. C., Breedveld, F. C., Smeets, T. J. M., Daha, M. R., Kluin, P. M., *et al.* (1996) Decrease in cellularity and expression of adhesion molecules by anti-tumor necrosis factor α monoclonal antibody treatment in patients with rheumatoid arthritis. *Arthritis Rheum.*, **39**, 1077.

111. Paleolog, E. M., Hunt, M., Elliott, M. J., Feldmann, M., Maini, R. N., and Woody, J. N. (1996) Deactivation of vascular endothelium by monoclonal anti-tumor necrosis factor α antibody in rheumatoid arthritis. *Arthritis Rheum.*, **39**, 1082.

112. Maini, R. N., Elliott, M. J., Brennan, F. M., Williams, R. O., Chu, C. Q., Paleolog, E., *et al.* (1995) Monoclonal anti-TNFα antibody as a probe of pathogenesis and therapy of rheumatoid disease. *Immunol. Revs*, **144**, 195.

113. Williams, R. O., Mason, L. J., Feldmann, M., and Maini, R. N. (1994) Synergy between anti-CD4 and anti-TNF in the amelioration of established collagen-induced arthritis. *Proc. Natl. Acad. Sci. USA*, **91**, 2762.

4 | Cytokines and the Th1/Th2 paradigm

SERGIO ROMAGNANI

1. Definition of Th1 and Th2 cells

In 1986, Tim Mosmann and Robert Coffman showed that the repeated stimulation of murine $CD4^+$ T helper (Th) lymphocytes with appropriate antigens results in the development of restricted and stereotyped patterns of cytokine production (Type 1 or Th1 and Type 2 or Th2). Murine Th1 cells produce interferon-gamma (IFN-γ), interleukin (IL) 2, and tumour necrosis factor (TNF)-β, and promote the production of IgG2a opsonizing and complement-fixing antibodies, macrophage activation, antibody-dependent cell cytotoxicity, and delayed-type hypersensitivity (1). For these reasons, Th1 cells have been considered as responsible for phagocyte-dependent protective host responses (2, 3) (Fig. 1). Th2 cells produce IL-4, IL-5, IL-6, IL-9, IL-10, and IL-13 and provide optimal help for humoral immune responses, including IgE and IgG1 isotype switching and mucosal immunity, through production of mast cell and eosinophil growth and differentiation, and facilitation to IgA synthesis. Moreover, some Th2-derived cytokines, such as IL-4, IL-10, and IL-13 inhibit several macrophage functions (1, 2). Therefore, Th2 cells may be considered as responsible for phagocyte-independent protective host responses (2, 3) (Fig. 1). In the absence of clear polarizing signals, $CD4^+$ T cell subsets with a less differentiated lymphokine profile than Th1 or Th2 cells, designated Th0, usually arise that mediate intermediate effects depending upon the ratio of lymphokines produced and the nature of the responding cells. Th0 cells probably represent a heterogeneous population of partially differentiated effector cells consisting of multiple discrete subsets that can secrete both Th1 and Th2 cytokines. The cytokine response at effector level can remain mixed or can be induced to further differentiate into the Th1 or Th2 pathway under the influence of signals received from the microenvironment (2). Some studies, however, have demonstrated heterogeneity of cytokine synthesis at the single cell level even in polarized Th1 and Th2 responses. Thus, another possibility is that cytokine profiles are largely random at the clonal level and that the exogenous signals, which appear to direct T cells to differentiate into Th1 or Th2 cells (see below), act by increasing the probability of expression of certain cytokine genes at the population level, rather than by activating the expression of a cassette of transcriptionally linked

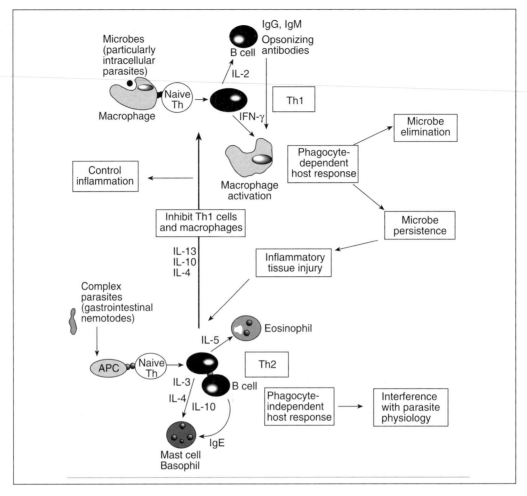

Fig. 1 Schematic view of the two polarized forms of the CD4$^+$-mediated effector response, based on the different profile of cytokine production. Th1-type cytokines mainly act via the activation of phagocytic cells and the production of opsonizing antibodies, thus being responsible for highly productive phagocyte-dependent responses against the majority of microbes. Th2-type cytokines favour eosinophil differentiation and activation, the growth of mast cells and basophils, as well as the production of large amounts of antibodies, including (IgE) by B cells. Th2-type cytokines also inhibit the Th1 development and several macrophage functions, thus providing phagocytic-independent protection against some gastrointestinal nematodes, and also acting as regulatory cells for Th1-mediated responses when they become dangerous for the host.

genes in the individual cell (4). The majority of authors now speak of Th2 (or Th2-like) cells as CD4$^+$ T cells that have been differentiated to produce IL-4 but not IFN-γ, and of Th1 (or Th1-like) cells as CD4$^+$ T cells that produce IFN-γ but not IL-4. Moreover, since Th2 cells can develop into IFN-γ-producing cells, whereas vigorously primed Th1 cells fail to develop into IL-4 producers, another appropriate basis on which to distinguish Th1 and Th2 effectors is that Th2 cells can produce IL-4 whereas Th1 cells cannot (5). Whatever definition is used, the Th1/Th2 model provides an interesting paradigm for the study and understanding of several pathophysiological processes and possibly for the development of novel immunotherapeutic strategies.

Evidence for the existence in humans of Th1 and Th2 cells similar to those described in mice was provided by establishing CD4$^+$ T cell clones specific for peculiar antigens (2). In general, however, human T cell clones have been found to exhibit a less restricted cytokine profile than murine T cells. IL-2, IL-6, IL-10, and IL-13 tend to segregate less clearly among human CD4$^+$ subsets than in the mouse. In addition, unlike mouse IL-10, human IL-10 inhibits the proliferative response and lymphokine production not only by Th1 but also by Th2 cells (2). As yet, no selective markers for Th1 or Th2 cells have been identified, although some surface molecules appear to be preferentially associated with the human Th1 or the Th2 phenotype. CD26, LAG-3, and the chemokine receptor CCR5 are preferentially expressed by Th1 cells (6–8), whereas CD30 and the chemokine receptors CCR3 and CCR4 preferentially associate with Th2 cells (8–10) (see also Chapter 5).

2. Nature of Th1/ Th2-polarizing signals

Several experimental models using limiting dilution analysis, immunization with oligopeptides, and bulk culture experiments with homogeneous population of cells from T cell receptor (TCR) transgenic mice have demonstrated the ability of a single Th cell precursor to differentiate to either a Th1 or Th2 phenotype. Using transgenic mice in which IL-4-producing cells expressed herpes simplex virus 1 thymidine kinase and could, therefore, be eliminated by ganciclovir, it was shown that activation of transgenic T cells in the presence of ganciclovir eliminates IL-4 and IFN-γ production (11). This very elegant experiment demonstrated that effector cells producing either IL-4 or IFN-γ have a common precursor, which expresses the IL-4 gene. The realization that Th1 and Th2 cells differentiate from a common pool of precursors allows questions to be asked about the factors that affect these differentiation pathways (Fig. 2).

2.1 Site of antigen presentation

The site of antigen presentation may play some role in determining the type of effector response. Inhaled ovalbumin (OVA) has a greater propensity than parenterally injected OVA to stimulate a Th2 response (12). Likewise, oral immunization with tetanus toxoid also selectively induces Th2 cells in mucosa-associated tissues (13). These findings may reflect unique populations of T cells or antigen-presenting cells (APCs) in the lung or gut, the influence of cytokines that are expressed at these sites, or other uncharacterized factors.

2.2 Type of APC

Attention has also been focused on the possibility that the type of response is dependent upon the nature of APC. However, no independent relationship between

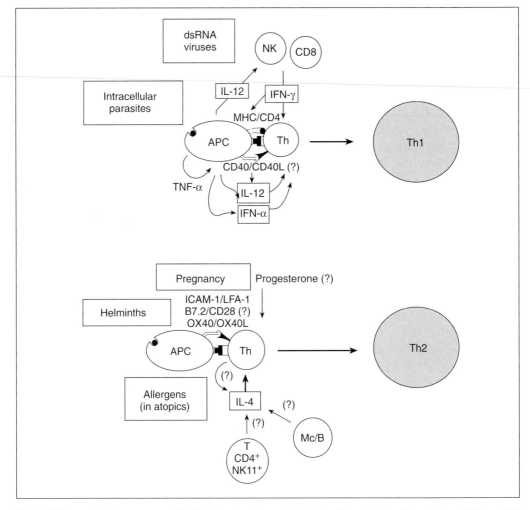

Fig. 2 Main mechanisms responsible for the development of naive Th cells into the polarized Th1 or Th2 phenotype. IL-12 and IFNs act as powerful Th1 inducers, whereas early IL-4 production favours the Th2 cell development. Progesterone during pregnancy may also favour, at fetal-placental level, a Th2 switch, which protects the fetal allotransplant against rejection mediated by the mother's Th1 cells.

type of APC and cytokine response has been definitely established. Mice immunized with antibodies that are focused onto B cells, macrophages, or dendritic cells all generate predominantly IgG1 responses, and macrophages, dendritic cells, and B cells can all induce either a Th1 or a Th2 response, given the proper cytokine environment (14).

2.3   Co-stimulatory molecules

Several co-stimulatory molecules present on the surface of APCs are involved in the T cell activation and function. Co-stimulatory molecules that have received considerable attention as possible regulatory signals for Th1/Th2 development are CD80/ CD86-CD28, CD4-Major histocompatability complex, CD54-CD11a/CD18, and OX40/OX40L. However, no definite conclusion on the critical role of these inter- actions in determining the pattern of Th differentiation has been achieved (15).

2.4 Properties of the immunogen

In general, it appears that corpuscolate immunogens more easily induce Th1 re- sponses than soluble antigens. The Th1-inducing activity of corpuscolate immuno- gens is probably related to their greater ability to induce IL-12 production by macro- phages that are responsible for phagocytosis. However, IL-12 can also be produced by dendritic cells (see below); moreover, some animal strains, as well as some humans, mount prevalent Th2 responses even in response to intracellular parasites that usually evoke Th1 responses because of a particular genetic background. The type of adjuvant is also important. CFA evokes Th1 responses, probably because the components of the mycobacterial cell wall induce IL-12 production by macrophages, with subsequent up-regulation of IFN-γ production (see below), whereas alum and acellular *Bordetella pertussis* toxin expand prevalent Th2 responses. Additional evidence on the role of immunogen in determining the type of Th cell development has been achieved by vaccination with plasmid vectors (naked DNA or poly- nucleotide vaccines). Recent studies suggest that immunization with naked plasmid DNA induce a Th1 immune response, which seems to be caused by the activity of unmethylated CpG dinucleotide (CpG ODN) sequences frequently present in the genome of viruses and bacteria, but not of vertebrates (16). These CpG ODN induce the release of high concentrations of IL-12, IFN-α, IFN-β, TNF-α, and IL-18 by macrophages or dendritic cells, which provide the optimal microenvironment for the development of Th1 responses (see below).

2.5 Dose of antigen

Several studies have suggested that differentiated effector CD4$^+$ T cells may produce different cytokines depending on the dose of antigen used. Low doses induced DTH by stimulating the development of Th1 cells, whereas higher doses induced antibody production because of the predominant production of Th2-type cytokines. However, these studies involved priming with complex antigens, such that different peptides might be presented at varying densities depending on the antigen dose. Exposure of CD4$^+$ T cells to low to medium doses of purified peptide promoted the development of Th1 cells, whereas increasing the dose of antigen resulted in the disappearance of IFN-γ and the development of IL-4-producing cells. At extremely low doses of antigen, IL-4 production was dominant over IFN-γ, regardless of the APCs used (17).

2.6 Peptide density and binding affinity

It is now clear that varying either the antigenic peptide or the MHC class II molecules can determine whether Th1- or Th2-like responses are obtained. High MHC class II peptide density on the APC surface favoured Th1-like responses, whereas low ligand densities favoured Th2-like responses. Likewise, by using a set of ligands with various class binding affinities but unchanged T cell specificity, it was shown that stimulation with the highest affinity ligand resulted in IFN-γ production. In contrast, ligands that demonstrated relatively lower MHC class II binding induced only IL-4 secretion (18). Taken all together, these findings suggest that the MHC binding affinity of antigenic determinants, which leads to differential interactions at the T cell–APC interface, can be crucial for the differential development of cytokine patterns in T cells (19).

2.7 Cytokines

There is general consensus that cytokines themselves are the major factors determining the differentiation of naive, and probably even memory, Th cells into the polarized Th1 or Th2 phenotype. Likewise, most studies in both mice and humans agree that IL-12 and IL-4 are the most important cytokines in directing the development of Th1 and Th2 cells, respectively.

2.7.1 Th1-inducing cytokines

The Th1-inducing effect of IL-12 was contemporarily and independently demonstrated in both mice and humans. In mice expressing an OVA-specific transgenic TCR, dendritic cells induced strong antigen-specific proliferation of CD4$^+$ T cells. However, addition of *Listeria* monocytogenes-activated macrophages, which produce high levels of IL-12, were essential for the development of Th1 cells that produce high levels of IFN-γ (20). In humans, the addition in peripheral blood mononuclear cells (PBMC) bulk culture of IL-12 shifted the development of Dermatophagoides group I (Der p I)-specific T cells from the Th0/Th2 to the Th1 profile. Conversely, the addition in PBMC bulk culture of anti-IL-12 antibody shifted the development of purified protein derivative (PPD)-specific T cells from the Th1 to the Th0 profile (21). Therefore, not only did IL-12 appear to be a factor able to favour Th1 differentiation, but the development of Th1 cells in the PBMC cultures that were stimulated with PPD were shown to depend, at least in part, on the production of endogenous IL-12 in culture. In this *in vitro* system, IL-12 not only induced strong augmentation in IFN-γ production, but also inhibited the development of IL-4-producing T cells. IL-12-driven development of murine Th1 cells is dependent on another cytokine, IFN-γ. When the endogenous IFN-γ that is present in primary T cell cultures was neutralized, IFN-α treatment augmented IL-12-induced effects on the inhibition of subsequent IL-4 production, even if it failed to significantly enhance IL-12 priming for subsequent IFN-γ production (22). The role of IFN-γ and IFN-α in favouring the *in vitro* development of Th1 cells has also been demonstrated in humans (23). Of note is that IFN-α seems to be capable of up-regulating the expression of the β chain of IL-12 receptor, thus favouring the response of T cells to this cytokine (24). Finally, IL-18 or IFN-γ-inducing factor (IGIF), can also contribute to the development of Th1 cells (25).

2.7.2 Th2-inducing cytokines

The addition of exogenous IL-4 to *in vitro* culture of naive, CD4$^+$ murine T cells in the presence of mitogens or antigens directs the naive CD4$^+$ T cells to develop into a population of effectors, which on restimulation produce high levels of IL-4, IL-5, IL-10, as well as IL-3 and granulocyte-macrophage colony-stimulating factor (GM-CSF) (Th2 pattern) (26). The effects of IL-4 include suppression of the development of IL-2 and IFN-γ secreting effectors, as well as the promotion of those of the Th2 pattern. Studies performed in mice that were transgenic for genes for the α and β chains of TCR specific for particular antigens, have confirmed that the presence of IL-4 during primary stimulation of naive CD4$^+$ T cells with mitogen or specific antigen plus APC results in the development of T cell populations that can secrete high levels of IL-4 upon restimulation (27). Moreover, nematode infection of mice lacking a functional IL-4 gene, as a result of functional inactivation of the gene by homologous recombination, resulted in much reduced production of Th2 cytokines IL-5, IL-9, and IL-10 from CD4$^+$ T cells when compared with wild-type control mice (28). However, even in IL-4 knock-out mice the Th2 response was not completely absent, suggesting that in spite of the IL-4's dominant role, factors other than IL-4 may also direct the Th2 development (28). CD4$^+$ populations producing high amounts of IL-4 and IL-5, in addition to IFN-γ and IL-2 (Th0 cells) or even IL-4 and/or IL-5 alone (Th2 cells), were also found to develop from human peripheral blood T cells stimulated with PPD (an antigen that usually favours the development of Th1-like cells) in the presence of exogenous IL-4 (29). IL-4, added in bulk culture before cloning, inhibited not only the differentiation of PPD-specific T cells into Th1-like clones, but also the development of their cytolytic potential. The depressive effect of IL-4 on the development of PPD-specific T cells with both Th1 profile and cytolytic potential was dependent on early addition of IL-4 (29).

Major candidates for IL-4 production at the onset of an immune response include FcεR1$^+$ cells, a specialized T cell subset known as CD4$^+$NK1.1$^+$ T cells, or the naive Th cell itself. Several types of FcεR1$^+$ non-T cells are able to produce IL-4. The IL-4-producing capacity of mouse non-T cells expands dramatically in *Nippostrongylus brasiliensis* infection and in association with anti-IgD injection (30), which suggests that these cells participate in lymphokine production in helminthic infections, and in other situations that are marked by striking elevations of serum IgE levels. These non-T/non-B cells represent the dominant source of IL-4 and IL-6 in the spleens of immunized animals. Exposing these cells to antigen-specific IgE or IgG *in vivo* (or *in vitro*) 'armed' them to release both IL-4 and IL-6 upon subsequent antigenic challenge (31). Accordingly, human mast cells, basophils, and eosinophils produce IL-4 in response to several secretagogues (32–34). However, it is unlikely that parasites or allergens would be able to cross-link their receptors prior to a specific immune response that had produced parasite-specific IgG and IgE antibodies. A way out of this dilemma may be a pathway of IL-4 secretion independent from FcεR cross-linking. Helminth products and some allergens may induce FcεR$^+$ cells to release IL-4 because of their proteolytic activity or via the induction of C5a. However, obvious mechanisms for FcεR-independent IL-4 production for the great majority of allergens or helminth components have not been identified yet. On the other hand, mast cell-deficient mice develop normal Th2 responses (35). Finally, in IL-4-deficient mice only those mice that are reconstituted with IL-4-producing T cells (but not with IL-4-producing non-T cells) produce antigen-specific IgE (36).

CD4$^+$NK1.1$^+$ T cells represent a specialized subset of T cells that are selected by the non-polymorphic MHC class I molecule, CD1 (37). The role of these cells in favouring the development of Th2 cells by providing IL-4 at the onset of an immune responses is suggested by several findings. First, splenic CD4$^+$NK1.1$^+$ cells are able to secrete large and transient amounts of IL-4 mRNA as soon as 90 minutes after intravenous injection of anti-CD3 antibody (38). Moreover, β$_2$-microglobulin(β$_2$M)-deficient mice and SJL mice, which contain a few or no CD4$^+$NK1.1$^+$ T cells, appear to be severely affected in their capacity for rapid induction of IL-4 mRNA in response to anti-CD3, as well as in their ability to synthesize IgE (39, 40). However, studies investigating *Leishmania major* infection and responses to other pathogens or antigens were unable to incriminate a role for these cells in Th2 responses (41). Finally, it is unlikely that all antigens that promote the differentiation of naive Th cells into the Th2 pathway are capable of activating CD4$^+$NK1.1$^+$, CD1-restricted T cells.

A final possibility is that the maturation of naive T cells into the Th2 pathway mainly depends upon the levels and kinetics of IL-4 production by naive T cells themselves at priming. This possibility was strongly supported by several findings obtained in both mice and humans. First, low intensity signalling of T cell receptor (TCR), such as that mediated by low peptide doses or by mutant peptides, led to secretion of low levels of IL-4 by murine naive T cells (42). Naive T cells, recently activated in the presence of co-stimulatory molecules-expressing fibroblasts (in the absence of outside influences from other cells), required two or more stimulation events to produce IL-4 and IL-5. This induction of Th2-type cytokine secretion was blocked by inhibiting IL-4 action, which suggests a role for endogenous IL-4 produced by the naive T cells themselves (43). Likewise, human CD45RA$^+$ (naive) adult peripheral blood T cells, as well as human neonatal T cells, were found to develop into IL-4-producing cells in the absence of any pre-existing source of IL-4 and despite the presence of anti-IL-4 antibodies (44). Thus, the maturation of naive T cells into the Th2 pathway may depend upon the levels and the kinetics of autocrine IL-4 production at priming. Obviously, when CD1-restricted antigens are expressed on APC, CD4$^+$NK1$^+$ T cells that are able to rapidly release high amounts of IL-4 may contribute to the development of the Th2 pathway. Recently, however, in IL-4 receptor-deficient mice an IL-4-independent pathway for CD4$^+$ T cell IL-4 production has been revealed (45).

2.8 Hormones

A role for hormones in promoting the differentiation of Th cells or in favouring the shifting of already differentiated Th cells from one to another cytokine profile has also been suggested. It has been reported that glucocorticoids, androgen steroids, and calcitriol can influence the cytokine secretion profile of Th cells (46–48). More recently, we found that progesterone favours the *in vitro* development of human Th cells, which produces Th2-type cytokines and promotes both IL-4 and leukaemia inhibitory factor (LIF) production in human T cells (49). Moreover, relaxin (another corpus luteum-derived hormone) was found to favour the development of IFN-γ-producing cells, without having any influence on IL-4 and IL-5 production, thus showing an opposite effect compared with progesterone (50). Therefore, increasing evidence is accumulating to suggest that hormones and peripherally activated prohormones may regulate Th1/Th2 balance.

2.9   Role of genetic background in Th1/Th2 development

From the above mentioned findings, it appears clearly that the type of antigen, antigen presentation, the type of adjuvant, and other environmental factors strongly influence Th1/Th2 development. However, strong evidence suggests the existence of striking differences in Th outcome depending on the genetic background of the host. The clearest example of genetic control of the Th1/Th2 responses is provided by murine *L. major* infection. Transgenic T cells from both B10.D2 and BALB/c backgrounds showed development toward either the Th1 or Th2 phenotype under the strong directing influence of IL-12 and IL-4, respectively. However, when T cells were activated *in vitro* under neutral conditions in which exogenous cytokines were not added, B10.D2-derived T cells acquired a significantly stronger Th1 phenotype than T cells from a BALB/c background, which is correspondent with *in vivo* Th responses to *L. major* in these strains (51). Another interesting example is provided by the different effect exerted by $HgCl_2$. In Brown Norway rats, $HgCl_2$ induces early and strong IL-4 expression and autoimmunity, whereas it does not affect IL-4 expression in Lewis rats who do not develop autoimmunity (52). Recently, it has been found that T cells from B10.D2 mice (a murine strain easily inducible to Th1 responses) have an intrinsically greater capacity to maintain IL-12 responsiveness under neutral conditions *in vitro* compared with T cells from BALB/c mice (a strain inducible to Th2 responses), allowing for a prolonged capacity to undergo IL-12-induced Th1 development (53). At present, however, the genetic mechanisms possibly controlling the differentiation of naive Th cells into the Th1 or Th2 phenotype remain largely unclear.

3.   Intracellular signalling for Th1/Th2 development

The appreciation that cytokines present in the microenvironment at the time of antigen presentation (IL-12/IFN and IL-4, respectively) play a critical role in favouring the differentiation of Th cells towards the Th1 or the Th2 profile has opened the question of the intracellular signalling involved in Th1/Th2 development. Cytokines exert their effects on cells by interacting with specific receptors expressed at the cell surface. This results in receptor homo- or heterodimerization, and triggering of intracellular signals (see also Chapter 2). One of the earliest signalling events is the activation of protein tyrosine kinase (PTKs). These kinases preferentially and constitutively associate with the intracellular domains of cytokine receptors and become activated following ligand-induced assembly of receptor subunits at the cell surface. After activation, receptor-associated PTKs phosphorylate several substrates critical for signal transduction, including specific tyrosine residues in the cytoplasmic domains of the receptors. One important component of cytokine signalling is the specific transcriptional activation of target genes, which is very rapid and does not require the synthesis of new proteins.

3.1 Intracellular signalling for Th2 development

The recognition that IL-4 expression during immune responses is critical for determining the development of Th2 cells has intensified interest in the molecular basis of its regulation. It is now clear that following the interaction between IL-4 and its receptor on a given cell the IL-4-induced signal transducer and activator of transcription (STAT) protein, initially designated IL-4 STAT and hereafter termed STAT6, is activated. STAT6 is related in primary amino acid sequence to STAT1, yet encoded by a different gene. The essential role of STAT6 in IL-4 signalling has clearly been demonstrated in STAT6-deficient mice. In these animals, T cells are indeed unable to develop into Th2 cells and the production of IgE and IgG_1 is virtually abolished (54, 55). However, there is as yet no direct evidence that STAT6 transactivates the IL-4 promoter in T cells or that the STAT6 site of the IL-4 promoter is required for promoter activity. On the other hand, other transcription factors of the nuclear factor of activated T cells (NF-AT) family are able to transactivate the IL-4 promoter, but they are expressed in both Th1 and Th2 cells (56). Thus it is unlikely that they can explain the Th2-specific expression of the IL-4 gene. NIP45 (NF-AT interacting protein) is another factor expressed in Th2 cells that appears to function as a potent co-activator of IL-4 gene transcription (57). However, its expression in Th1 cells is unclear. By contrast, the proto-oncogene c-*maf* appears to be selectively expressed in Th2 clones and it is induced during Th2, but not Th1 differentiation (58). Moreover, its activity appears to be specific of the IL-4 promoter, inasmuch as it is able to transactivate the promoters of other Th2 cytokine genes such as IL-5 or IL-10. A transcription factor that may be more widely involved in the induction and maintenance of the Th2 pattern of cytokine secretion is GATA-3 (59). GATA-3 is expressed in both immature and mature T cells, but is selectively suppressed during Th1, but not Th2 differentiation (59). Thus, in contrast to c-*maf*, which appears to be IL-4-specific, GATA-3 may function as a more general regulator of Th2 cytokine expression (60) (Fig. 3).

3.2 Intracellular signalling for Th1 development

Since IL-12 directly influences T cells for the induction of Th1 differentiation, signals delivered by IL-12 at the T cell surface would be an important step in inducing Th1

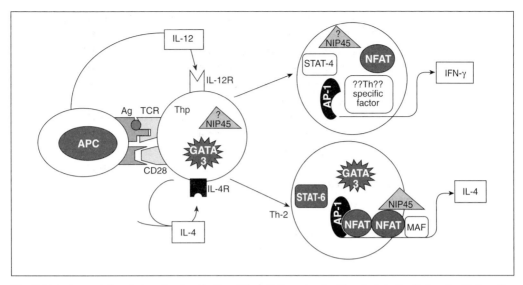

Fig. 3 Main transcription factors involved in the differentiation of naive Th cells into the Th1 or the Th2 profile. This field is now under continuous and rapid evolution.

development from naive T cells. Treatment of murine TCR-transgenic T cells with IL-12 produced an electromobility shift assay (EMSA) complex that contained STAT3 and STAT4, and whose mobility was distinct from that of complexes induced by IL-2 and IL-4, which contain STAT5 and STAT6, respectively (61). Interestingly, activation of EMSA complexes containing STAT3 and STAT4, as well as STAT4 phosphorylation, are evident in early Th1 cells. Th2 cells did not respond to IL-12, although maintaining the capacity to respond to several other cytokines, including IL-2, IL-4, IFN-γ (involving activity of JAK1 and JAK2), and IFN-α (involving activity of JAK1 and TYK2). The lack of STAT4 phosphorylation in Th2 cells was not caused by a lack of STAT4 protein, since Th2 cells expressed similar levels of STAT3 and STAT4 compared with Th1 cells, but in the failure to phosphorylate JAK2, STAT3, and STAT4 (62). This implies that the defect in Th2 cells for IL-12 responses is upstream of the STATs and the kinases, and may involve an unidentified component of the IL-12 receptor itself. In human T cells, IL-12 induces tyrosine phosphorylation of JAK2 and TYK2, whereas JAK1 and JAK3, which are phosphorylated in response to IL-2, are not phosphorylated after IL-12 treatment (62). More importantly, IL-12 induces tyrosine phosphorylation and activation of STAT4-, but not of STAT1- or STAT2-containing complexes, which bind to the GRR DNA sequence (63). In the presence of IL-12 and the absence of IL-4, c-*maf* expression is not up-regulated and GATA-3 expression is extinguished (60). Based on the current model, the lack of these two signals contributes to Th1 commitment and the absence of IL-4 expression. Whether a Th1 specific factor similar to GATA-3 for Th2 cells exists, remains to be established (Fig. 3).

4. Cross-regulatory activity of Th1 and Th2 cytokines

It is well established that some lymphokines produced by Th1 and Th2 cells can exert mutual regulatory interactions (see also Chapter 3). In particular, IL-4 and IFN-γ, the principal products of Th2 and Th1 cells, respectively, often oppose one another's actions. IL-10, which is mainly produced by Th2 cells but also by macrophages and B cells, exerts a negative regulatory effect on some Th1 cell functions. By contrast, the Th1-derived cytokine IFN-γ, as well as IL-12 and IFN-α, which are not a product of Th1 cells but act as Th1-inducing cytokines, down-regulate the function of Th2 cells. There is considerable evidence that IL-4 prevents priming of naive Th cells to become IFN-γ producers. In fact, IL-4 is able to strikingly inhibit IL-2 production by naive T cells in response to anti-CD3 antibody and APCs or, when transgenic T cells are used, to antigen and APCs (64). In contrast, IL-4 has little effect on IL-2 or IFN-γ production by established T cell lines or by T cells that have been already primed *in vitro*. The suppressive effects of IL-4 on priming for IFN-γ production are markedly diminished in the presence of IL-12. Thus IL-12 not only enhances the priming of naive Th cells to become IFN-γ producers, it also diminishes their sensitivity to inhibition by IL-4 (65).

IL-10, another cytokine selectively produced by murine Th2 cells, but common to all types of human Th cells, inhibits lymphokine production by Th1, but not by Th2 clones (66). This inhibition is somewhat selective in that secretion of IFN-γ and IL-3, lymphokines secreted mainly at later times after stimulation, is substantially inhibited, whereas secretion of GM-CSF and TNF-β, lymphokines secreted mainly at early times, is inhibited slightly if at all. The effect on IL-2 production often is less that the effect on IFN-γ production. IL-10 also inhibits the IFN-γ-mediated positive feedback stimulation of IL-12 production by macrophages. By suppressing co-stimulatory molecules such as TNF-α and IL-1, IL-10 inhibits the stimulatory effects of IL-12 on T and NK cells, which in turn results in down-regulation of Th1 cell differentiation (67).

The addition of IFN-γ to cultures that contain optimal amounts of IL-4 failed to inhibit the priming of CD4$^+$ T cells, obtained from TCR transgenic mice, to develop into IL-4-producing T cells. However, when suboptimal concentrations of IL-4 were used for priming, IFN-γ caused a significant decrease in the amount of IL-4 produced after restimulation (68). Although not absolute, the inhibitory effect of IFN-γ on proliferation of murine Th2 cells is significant and seems to be sufficient to limit the clonal expansion of such cells. Although IFN-γ does not inhibit lymphokine secretion by stimulated Th2 cells, IFN-γ does inhibit many of the agonist effects of those secreted cytokines. For example, IFN-γ inhibits proliferation of murine bone marrow cells stimulated with IL-3, IL-4, or GM-CSF, as well as IL-4-dependent B cell differentiation (69, 70). Like IFN-γ, IFN-α plays a negative regulatory role in the development of Th2 cells. In particular, in the mouse, IFN-α was found to be able to suppress increases in the level of splenic IL-4 mRNA induced by either treatment with anti-IgD antibody or infection with *Nippostrongylus brasiliensis*, whereas in both conditions the levels of IFN-γ mRNA were increased (71). In the absence of IFN-γ, IFN-α augments IL-12 effects on inhibition of subsequent IL-4 production rather than enhancing IL-12 priming for subsequent IFN-γ production (22). The regulatory activity of IFN-α has also been demonstrated in the human system. IFN-α inhibited the development of allergen-specific T cells into Th2-like cells (23). This effect is probably related to a selective antiproliferative activity on T cells planned to the production of Th2 cytokines, inasmuch as addition of IFN-α in bulk culture before cloning resulted in the development of allergen-specific T cell clones showing not only a different cytokine profile (Th1 or Th0 versus Th2), but also different TCR repertoire and peptide reactivity (72).

In addition to its well known ability in priming Th cells to produce IFN-γ and, therefore, to develop into Th1 cells, IL-12 also inhibits the differentiation of T cells into IL-4-secreting cells, i.e. into Th2 cells. This inhibitory effect is in part a direct effect, in part is mediated by its activity on APCs, and in part by its stimulation of IFN-γ synthesis (73). IL-12 is not capable, however, of suppressing a Th2 cell recall response. Interestingly, IL-12 may also limit its own effects by inducing the production of a cytokine, such as IL-10, that down-regulates both IL-12 production and IL-12-induced IFN-γ production. When IL-12 and IL-4 are present together, paradoxically IL-4 suppresses the IL-12-induced priming for IL-10 production, but only minimally decreases the priming for high IFN-γ production. The *in vivo* effects of IL-12, however, appear to be, to a large extent, IFN-γ-dependent (74).

5. The Th1/Th2 paradigm in human diseases

The Th1/Th2 paradigm has been found of great usefulness, not only in unravelling the mechanisms of protection against different biological agents, but also in better understanding the pathogenic mechanisms underlying several immunopathogenic disorders.

5.1 Transplantation rejection and tolerance

Several studies suggest that some Th1-related effector mechanisms, such as DTH and CTL activity, play a central role in acute allograft rejection. Proteins and/or transcripts for intragraft IL-2, IFN-γ, and the CTL-specific marker, granzyme B, have consistently been detected in rejecting allografts. IFN-γ is believed to recruit macrophages into the graft, cause macrophage activation, and enhance CTL activation (75, 76). Cytokines are probably involved in the genesis of the main anatomopathological lesions characteristic of allograft rejection. It has been shown that CD4[+] and CD8[+] T cell clones derived from rejected kidney grafts expressed a predominant Th1 lymphokine profile in comparison with T cell clones derived from the peripheral blood, and that Th1 cytokines (particularly IFN-γ) are prominently expressed in human renal allografts prior to or during rejection (77).

Transplantation tolerance can be defined as an inability of the organ graft recipient to express a graft destructive immune response. The production of Th2-type cytokines may be central to the induction and maintenance of allograft tolerance. Several *in vivo* studies that examined the pattern of cytokine expression during tolerance induction have consistently shown a dramatic decrease in the expression of IL-2 and IFN-γ, whereas increased levels of IL-4 and IL-10 transcripts are manifest. Thus, although its validity is not yet proved, the Th1/Th2 paradigm may represent the basis for understanding the mechanisms of rejection and tolerance in transplantation (75, 76).

5.2 Successful pregnancy and unexplained recurrent abortions

Despite the fact that the embryo, because of the presence of paternal MHC antigens, is alike to an allograft and therefore represents a potential target for the maternal immune system, it is usually not rejected until the time of delivery. This means that maternal immunological recognition of the developing fetal semi-allograft is a complex process that allows fetal survival and growth to occur (immunological tolerance of the conceptus). Lin *et al.* (78) have reported that the Th2-type cytokines IL-4, IL-5, and IL-10 are detectable at the materno-fetal interface during all period of gestation, whereas IFN-γ production is transient, being detectable only in the first period. Based on these findings, the existence of a bidirectional interaction between the maternal immune system and the reproductive system during pregnancy was hypothesized (79). We recently investigated decidual T cells of pregnant women during the first period of gestation (from 8–12 weeks) by comparing the cytokine profile of CD4$^+$ and CD8$^+$ T cell clones established from both decidua and blood of women with unexplained recurrent abortion (URA) or voluntary abortion (normal gestation), as well as from endometrium and blood of non-pregnant women. Interestingly, although the majority of T cell clones from the three groups studied showed a Th0-like profile, a significantly higher number of Th1-like T cell clones were generated from the decidua of women suffering from URA (156). Furthermore, T cell clones generated from the decidua of women with URA produced significantly lower IL-4, IL-10, and LIF concentrations than T cell clones generated from either the decidua of voluntary abortion or the endometrium of non-pregnant women (M.-P. Piccinni *et al.*, unpublished data). These results suggest that local production of IL-4 may be important for the maintenance of pregnancy, whereas its reduced production can compromise pregnancy.

5.3 Allergic disorders

Based on their cytokine secretion profile and related functions, Th2 cells appear to be excellent candidates to explain the involvement of the mast cell/eosinophil/IgE-producing B cell triad in the pathogenesis of allergy (80). In the last few years, strong evidence that supports the pathogenic role of allergen-reactive Th2 cells in allergic disorders has accumulated. First, allergen-specific CD4$^+$ T cell clones generated from the peripheral blood of atopic subjects produce IL-4 and IL-5, but little or no IFN-γ, whereas T cell clones specific for bacterial antigens generated from the same donors produce high IFN-γ and low or no IL-4 and IL-5 (81, 82). Most importantly, the presence of activated T cells exhibiting a Th2-like profile has been demonstrated in target organs of patients suffering from allergic disorders, such as vernal conjunctivitis, allergic rhinitis and asthma, and atopic dermatitis.

The first demonstration that Th2-like T cells can accumulate in tissues from allergic patients was provided by generating T cell clones from the conjunctiva of children suffering from vernal conjunctivitis, a condition characterized by papillary hypertrophy of the conjunctiva with eosinophil and lymphocyte infiltration and associated high serum IgE. T cell clones generated from the conjunctival tissue of children with vernal conjunctivitis were mostly CD4+, produced IL-4 but little IFN-γ, and supported IgE synthesis by B cells *in vitro* (83).

Nasal allergen provocation, followed by nasal mucosa biopsy 24 hours after allergen challenge, in patients with allergic rhinitis is associated with a cellular infiltrate where CD4+ T cells and eosinophils are prevalent (84). By *in situ* hybridization and immunohistochemistry, an increase in cytokine mRNA expression for IL-3, IL-4, IL-5, and GM-CSF was found, which correlated with the degree of local eosinophilia (84). Accordingly, CD4+ T cell clones generated from the nasal mucosa after allergen challenge, many of which were specific for the allergen used for challenge, exhibited a Th2 profile (85).

Asthma in humans is a complex disorder characterized by intermittent, reversible airway obstruction, and by airway hyperresponsiveness and inflammation. Asthma may be divided into allergic (extrinsic) and non-allergic (intrinsic) forms. Both allergic and intrinsic asthma are characterized by infiltration of the bronchial mucosa with large numbers of activated eosinophils and the presence of elevated concentrations of eosinophil-derived proteins. Using an *in situ* hybridization technique, cells showing mRNA for Th2, but not Th1, lymphokines were found at the site of late phase reactions in skin biopsies from atopic patients (86), and in mucosal bronchial biopsies or BAL of patients with atopic asthma (87, 88). The majority of IL-4 and IL-5 mRNA in BAL cells from asthmatic subjects was shown to be associated with CD2+ T cells (88). T cell clones were also generated from biopsy specimens of bronchial mucosa of patients with grass pollen-induced asthma or rhinitis, taken 48 hours after positive bronchial provocation test with the relevant allergen. Proportions that ranged from 14–22% of CD4+ clones derived from stimulated mucosa of grass-allergic patients were specific for grass allergens, and most of them exhibited a definite Th2 profile and induced IgE production by autologous B cells in the presence of the specific allergen (85). In contrast, none of the T cell clones, derived from the bronchial mucosa of patients with TDI-induced asthma at 48 hours after positive bronchial provocation test with TDI, were specific for grass allergens and the majority of them were CD8+ T cells producing IFN-γ and IL-2 or IFN-γ, IL-2, and IL-5, but not IL-4 (85). Accordingly, allergen inhalation challenge resulted in the activation of CD4+ T cells, increased Th2-type cytokine mRNA expression, and eosinophil recruitment in BAL of patients with atopic asthma (89, 90).

Atopic dermatitis (AD) is the cutaneous disorder of atopic allergy caused by chronic skin inflammation, which results from the exposure to foreign proteins. The majority of patients with AD (> 80%) have positive intracutaneous skin test reactions to one or more environmental allergens and elevated serum IgE levels, which represent antibodies specific to the allergens concerned. However, the relationship between allergy and the pathogenesis of the skin lesions in AD is still unclear. By using the RT–PCR technique, spontaneous mRNA IL-4 expression was found in peripheral blood mononuclear cells from AD patient (91), whereas IL-13, another IgE-switching cytokine, was not produced (92).

High proportions of Dp-specific Th2-like CD4$^+$ T cell clones were obtained from the skin lesions of patients with AD, indicating accumulation or expansion of these T cells in lesional skin. Interestingly, Dp-specific Th2-like clones were also derived from biopsy specimens of intact skin taken after contact challenge with Dp, which suggests that percutaneous sensitization to aeroallergens may play a role in the induction of skin lesions in patients with AD (93–95). More recent data partially confirm both the Der p-specificity and the Th2-type profile of high proportions of T cell clones derived from Der p-induced patch test lesions of AD patients. However, the majority of T cell clones derived from the lesional skin of patients with AD had a mixed (Th0-like) phenotype and only a minority of them were specific for Dp. Interestingly, Der p-specific T cell clones generated from the skin of AD patients usually have a more polarized Th2 cytokine profile than those obtained from the peripheral blood of the same patients, and consistently produce IL-10 in addition to IL-4 and IL-5 (96). In some patients, most T cell clones derived from lesional skin exhibited high IFN-γ production and a few of them appeared to be specific for bacterial antigens. More importantly, the presence of both Dp-specific and Th2-like cells in the skin of AD patients did not correlate with the presence in the serum of Dp-specific IgE antibody or the serum levels of IgE protein (97). Finally, by assessing the kinetics of T cell-derived cytokine production, it was found that in the initiation phase (APT skin site) IL-4 production is predominant over IFN-γ production. In the late and chronic phase (lesional skin) the situation may be reversed and IFN production predominates over IL-4 production (98). Taken together, these data suggest that Th2-like responses against Dp or other allergens at skin level are involved in the initiation of skin lesions, but there is a subsequent influx of Th1 cells, which may be responsible for the aggravation of the inflammation.

5.4   Autoimmune disorders

Several findings indicate that autoimmune diseases develop as a result of abnormalities in the immune response mediated by activated T cells and T cell-derived lymphokines. Strong evidence derived from studies in animal models and investigations in human disease suggests that Th1-type lymphokines are involved in the genesis of organ-specific autoimmune diseases. In contrast, a less restricted lymphokine pattern is emerging from experimental studies on systemic autoimmune disease. Accordingly, data available from human disease studies are in favour of a prevalent Th1 lymphokine profile in target organs of patients with organ-specific autoimmunity, such as Hashimoto's thyroiditis, multiple sclerosis (MS), and Type 1 insulin-dependent diabetes mellitus (IDDM), whereas a less restricted lymphokine pattern is detectable in patients with rheumatoid arthritis (see also Chapter 3). By contrast, a prevalent Th2 pattern can be observed in patients with systemic sclerosis (SSc) (99).

5.4.1 Autoimmune thyroid diseases

Although the precise aetiology of HT remains largely unknown, there is general consensus that infiltrating T lymphocytes play an essential role in the destruction of target organs. Several studies have shown that T cells from lymphocytic thyroid in-filtrates of patients with HT or GD exhibit both a restricted Th1 lymphokine profile, with production of high TNF-α and IFN-γ concentrations, and strong cytolytic potential (100–102). A prevalent Th1-like cytokine profile was also observed by RT–PCR in the thyroid gland of patients with HT. A homogeneous Th1 phenotype was also observed in CD4$^+$ T cell clones derived from retro-orbital infiltrates of patients with Graves' ophthalmopathy (103). Similar results were obtained by analysing cytokine gene expression by RT–PCR in intrathyroidal lymphocytes from six patients with GD. IL-2, IFN-γ, and TNF-α, but not IL-4, mRNA were found, suggesting clear predomination of Th1 responses (104). In contrast, by using the same technique, other authors found a more heterogeneous cytokine profile in both the thyroid gland and retro-orbital infiltrates of patients with GD (105–108). The reason for these dis-crepancies may be explained with the different cytokine profiles of thyroid antigen-specific T cells. Indeed, the majority of thyroid peroxidase (TPO)-specific clones showed a Th1 profile, whereas the cytokine profile of clones specific for thyroid-stimulating hormone receptor (TSHR) were more characteristic of Th0 or Th2 cells (109). Thus, it is possible that the thyroid microenvironment allows the expansion not only of autoreactive T cells able to release IFN-γ, but also of other clones that secrete IL-4 and IL-10, and, therefore, are more active in promoting the synthesis of pathogenic autoantibodies.

5.4.2 Multiple sclerosis (MS)

Several studies in human disease suggest a role for TNF-α and IFN-γ in the pathogenesis of MS. First, high levels of TNF-α are present in the cerebrospinal fluid of patients with chronic progressive MS. Determination of TNF-α in both plasma and cerebrospinal fluid may predict relapses in MS patients (110). By using RT–PCR, high levels of TNF-α mRNA were also found in PBMC from MS patients and increased TNF-α expression appeared to precede relapses by four weeks in patients with relapsing remitting MS, whereas IL-10 and TGF-β expression was decreased at the same time (111, 112). Most clones derived from both PBMC and cerebrospinal fluid of patients with MS showed a Th1 profile (113, 114). Increased numbers of IFN-γ-secreting T cells, which produced IFN-γ upon activation by several myelin antigens and several MBP peptides were found in the cerebrospinal fluid of MS patients (115). T cell clones specific for PLP peptides generated from MS during an acute attack showed a clear-cut Th1 profile, whereas during remission in the same patients a more heterogeneous cytokine profile was found. These clones produced levels of both IL-10 and TGF-β significantly higher than those of clones isolated during acute attacks (116). Using immunohistochemistry, both TNF-α and TNF-β were identified in acute and chronic active MS lesions, but not in spleens or PBMCs from MS patients. TNF-β expression was associated with T lymphocytes, whereas TNF-α was associated with astrocytes in all areas of the lesion (117). By using semi-quantitative RT–PCR and immunohistochemistry, the co-stimulatory molecules B7-1, B7-2, and the cytokine IL-12p40 were found to be up-regulated in acute MS plaques from early disease cases. The differences in cytokine mRNA expression were specific for IL-12p40, whereas no differences were observed for other cytokines, suggesting that an early event in the initiation of MS involved up-regulation of co-stimulatory molecules and IL-12 (118). These probably represent the conditions that maximally stimulate T cell activation and induce Th1-type immune responses. Finally, the antagonistic effects of treatments with IFN-γ and IFN-β on the course of the clinical outcome of the disease provide further indirect support for a pathogenic role of Th1 cells in MS. In some clinical trials, it has been shown that IFN-γ administration induces relapses of MS (119), probably because it up-regulates MHC class II expression by microglial cells and APC-like macrophages and stimulates the secretion of potentially myelino-toxic mediators, such as TNF-α from macrophages. In contrast, fibroblast- and leukocyte-derived IFN-β, which inhibits IFN-γ-induced MHC class II expression, has been shown to reduce the rate of relapse in MS (120). In conclusion, although studies of the T cell immune response in MS are not yet so clear as studies of experimental autoimmune encephalitis (EAE), there is convincing evidence to suggest that Th1 cells play a critical pathogenic role in this disease.

5.4.3 Type 1 or insulin-dependent diabetes mellitus

The development of insulin-dependent diabetes mellitus (IDDM) in humans and in the spontaneous animal models of the BB rat and the non-obese diabetic (NOD) mouse, is the result of a cellular autoimmune process that selectively destroys the pancreatic islet beta cells. Evidence suggesting a role for Th1 cells in the pathogenesis of IDDM is supported by studies on BB and NOD mice. In BB mice, both IL-12 p40 chain and IFN-γ mRNA are present in the inflammatory lesions, whereas mRNA for IL-2 and IL-4 is minimal or undetectable (121). All NOD mice spontaneously develop insulitis early in life (between two and four weeks of age), but it is not until 10–20 weeks later that this insulitis progresses to diabetes in about 80% of female mice and in only 20% of male mice. There is extensive infiltration of the islets by CD4$^+$ and CD8$^+$ T cells, B cells, and macrophages, as occurs in human IDDM. Although it is possible that CD8$^+$ T cells can cause diabetes in NOD mice in the absence of CD4$^+$ T cells (122), Th1 cells play a major role in the induction of diabetes. IFN-γ-producing Th1-like clones cause IDDM after transfer into neonatal NOD mice, and the transfer of T cell clones that secrete IFN-γ and IL-2 is able to precipitate diabetes in recipient NOD mice. By contrast, Th2 clones are unable to transfer diabetes. An injection of IL-12 also induces massive Th1 cell infiltration in the pancreatic islets and accelerates IDDM development in NOD mice, although intermittent IL-12 administration was, surprisingly, found to delay IDDM development (123). Moreover, IL-12 p40 knock-out mice on a NOD background, which produce no or very low IFN-γ, showed almost no insulitis at any age tested and their spontaneous IDDM incidence was very low.

There are a few contributions on the characterization of the inflammatory cell infiltrate in the insulitis lesion in human Type 1 diabetes. Such studies have had to be largely confined to autopsies on patients who died of recent onset diabetes where the pancreatic tissue had been formalin-fixed. Thus, the combined problems of autolysis, caused by the interval between death and autopsy, and protein denaturation, caused by fixation, have severely limited the knowledge in this field. However, it appears that the cell infiltrate is composed of both lymphocytes and macrophages (ratio 7–9:1) and that high proportions of lymphocytes (about 40%) contain IFN-γ (124). Autoreactive T cell clones generated from newly onset patients with IDDM exhibited a predominant Th1 profile, whereas those derived from a pre-diabetic patient were prevalently Th2 (125).

5.4.4 Systemic sclerosis

Systemic sclerosis (SSc) is a disorder characterized by inflammatory, vascular, and fibrotic changes of the skin (scleroderma) and a variety of internal organs, such as the gastrointestinal tract, lungs, heart, and kidney. In the skin, a thin epidermis overlies compact bundles of collagen that lie parallel to the epidermis. Increased numbers of T cells are present in skin lesions, as well as in other organs in the early stages of the disease. Several soluble factors secreted by T cells or other cells of the immune system may modulate fibrosis or promote vascular damage in SSc. For example, IL-4 induces human fibroblasts to synthesize elevated levels of extracellular matrix proteins, as well as to stimulate the growth of subconfluent fibroblasts and induce chemotaxis of these cells (126–128). Interestingly, PBMC from patients with SSc produce higher amounts of IL-2 and IL-4 than controls (129). Recently, we generated T cell clones from the skin cellular infiltrates of SSc patients, which were mostly Th2 (130). Accordingly, analysis by *in situ* hybridization and immunohistochemistry showed that large numbers of CD4+ T cells present in the perivascular infiltrates of skin biopsy specimens expressed mRNA for IL-4 but not for IFN-γ, and were CD30+ (130). High levels of sCD30 were also found in the serum of most SSc patients, especially those showing active disease (130). These data strongly suggest a predominant activation of Th2 cells in SSc and support the view that abnormal and persistent IL-4 production by the activated Th2 cells may play an important role in the genesis of fibrosis and, therefore, in the pathogenesis of the disease.

5.5 Chronic inflammatory gastrointestinal disorders

5.5.1 *Helicobacter pilori*-related gastroduodenal pathologies

Colonization of the mucosa of the stomach and the duodenum by *Helicobacter pilori* (Hp) is thought to be the major cause of acute and chronic gastroduodenal pathologies in humans. These include gastric and duodenal ulcer, as well as gastric carcinoma and lymphoma of the mucosa associated lymphoid tissue (MALT). The great majority of patients with duodenal ulcer disease harbour Hp infection in the gastric antrum and exhibit an associated chronic antral gastritis, which is characterized by a mucosal infiltrate of polymorphonuclear and mononuclear leukocytes.

Very recently, we have analysed the pattern of cytokines that are produced by the immunologically active cells within the mucosa from antral biopsies of Hp-infected patients with duodenal ulcer and Hp-negative dyspeptic controls by using RT–PCR and immunohistochemistry. T cell clones were also generated from parallel samples of the antral mucosa of the same Hp-infected patients and assessed for their re-activity with Hp antigens, profile of cytokine production, and effector functions. Antral biopsies from all Hp-infected patients with duodenal ulcer showed IL-12, IFN-γ, TNF-α, and IL-12, but not IL-4, mRNA expression, as well as the presence of IFN-γ-containing CD4$^+$ T cells, whereas neither cytokine mRNA signal nor cytokine-containing mononuclear cells were found in the mucosa of Hp-negative controls (131). When assayed for responsiveness to a Hp lysate, 24 of 163 CD4$^+$ clones (15%) exhibited marked proliferation in response to Hp lysate. 11 clones reacted with Cag-A, two with Vac-A, and one with Ure-B. Upon stimulation with the specific antigen, the great majority of Hp-reactive clones (20/24) produced IFN-γ but not IL-4 or IL-5 (Th1-like), whereas four clones produced both IFN-γ and IL-4 or IL-5 (Th0-like). Under the same experimental conditions, all Hp-reactive clones produced high con-centrations of TNF-α (131). These results demonstrate that Hp-specific T cells show-ing Th1-like profile of cytokine production and related effector functions are present in the gastric antral mucosa of patients with peptic ulcer. These cells may play a role in the pathogenesis of both peptic ulcer and gastric malignancies associated with Hp infection.

5.5.2 Chronic inflammatory bowel diseases

The idiopathic inflammatory bowel diseases (IBD) comprise a spectrum of disorders that are marked by the presence of chronic inflammation of the gastrointestinal tract that cannot be ascribed to a specific pathogen. At one end of the spectrum is ulcer-ative colitis (UC), a disease that affects the large bowel exclusively, at the other end of the spectrum is Crohn's disease (CD), that in contrast to ulcerative colitis can affect any part of the alimentary canal, from the mouth to the rectum, but most commonly involves the terminal ileum and the ascending colon. The aetiology of both UC and CD is unknown. Insults from microbial agents have been suspected, but there is no definite proof for the infective theory. Much circumstantial evidence, however, supports the concept that immunological mechanisms are responsible for the pathogenesis of both UC and CD.

In the last few years, an impressive number of different model of IBD in mice and rats have been described. An IBD-like picture develops in mice with alterations in T cell subpopulations and T cell selection (TCR α chain-deficient mice, TCR β chain-deficient mice, MHC class II-deficient mice) (157), in mice with targeted disruption of cytokine genes (IL-2-, IL-10-, and TGF-β-deficient mice) (132–135), in mice lacking signalling proteins (G protein subunit Gα2-deficient mice) (136), as well as in mice subject to rectal application of the hapten reagent 2,4,6-trinitrobenzene sulfonic acid (TNBS) (137). Data from a number of these models indicate that a subpopulation of T cells plays a critical role in the normal regulation of intestinal immune responses. Moreover, evidence is accumulating to suggest that cytokines are involved in the pathogenic and regulatory pathways. The Th1-derived cytokines IFN-γ and TNF-α appear to be responsible for the pathogenesis of colitis in SCID mice restored with memory CD4$^+$ T cells, as disease was prevented by the administration of anti-IFN-γ antibody and significantly reduced in severity by anti-TNF antibody (138). Moreover, antibodies to IL-12 abrogated established experimental colitis that was induced in mice by rectal application of the hapten reagent TNBS (137). Taken together, these data suggest a pivotal role of IL-12 and Th1 cytokines in the induction of murine chronic intestinal inflammation.

Increased levels of IL-1, IL-6, IL-8, TNF-α, as well as of IL-2 and IL-2 receptor, have been found in the intestinal mucosa and/or serum of patients with IBD, suggesting activation of macrophages and perhaps of Th1-like cells. Moreover, increased expression of mRNA for IL-2, IFN-γ, and IL-10 are present in the mucosa of patients with active CD, whereas IL-4 mRNA expression was frequently below the detection limits (139). In recent studies, it was found that T cell clones generated from the colonic mucosa of patients with CD are able to produce higher levels of IFN-γ, but significantly reduced levels of IL-4 and IL-5, in comparison with T cell clones derived from the colonic mucosa of patients with non-inflammatory bowel disorders (139, 140). In contrast, no differences in the production of IL-10 between the two groups of clones were found, suggesting that a defect in the regulatory role of IL-4 rather than of IL-10 may be operating in human IBD. IL-10, however, is not a selective Th2 product in human beings, but is also produced by Th1 and Th0 cells and even more by B cells and macrophages. Thus, a possible concomitant defect in IL-10 production by non-T cells from IBD patients cannot be excluded. More recently, the possible pathogenic role of Th1-dominated responses was further supported by the results of immunohistochemical studies. High numbers of activated CD4$^+$ T cells showing CD26, LAG-3, and IFN-γ reactivity, as well as of IL-12-containing macrophages, were found to infiltrate the muscularis of gut from patients with Crohn's disease, but not of controls. In addition, culturing IL-2-conditioned T cells from the mucosa of CD patients in the presence of anti-IL-12 antibody had an inhibitory effect on the development of IFN-γ producing T cells, suggesting that constitutive IL-12 production plays a critical role in the development of Th1 cells at intestinal level (139, 140).

5.6 Progression of HIV infection

The possible influence of cytokines produced by Th1 and Th2 cells in the progression of HIV infection has also been matter of intense debate. Based on the demonstration that Th1 responses are in general more protective than Th2 responses in the defence against intracellular pathogens, it has been suggested that a switch from the Th1 to the Th2 cytokine profile occurs and that this is a critical step in the progression towards full-blown AIDS (141). While the hypothesis of a Th1/Th2 switch in HIV infection could not be supported by the results of other studies (142, 143), we and others found that Th0 and Th2 clones support HIV-1 replication better than Th1 clones (143–145). Recently, we investigated the possible regulatory role of Th1 and Th2 cytokines on the expression of CXCR4 (a co-receptor for the entry into the cell of T-tropic HIV-1 strains) (146), as well as on HIV expression by using *in vitro* models of both secondary and primary immune responses.

Antigen-specific memory CD4$^+$ T cells infected with a T-tropic HIV-1 strain showed significantly higher CXCR4 and HIV-1 expression in T helper 0 or 2 (Th0/2)-oriented in comparison with T helper 1 (Th1)-oriented responses. Similarly, in naive CD4$^+$ T cells activated in the presence of IL-4 or IL-12 and infected with the same T-tropic strain, IL-4 up-regulated whereas IL-12 down-regulated both CXCR4 and HIV-1 expression. The down-regulatory effect of IL-12 on CXCR4 expression was found to be dependent on the induction of IFN-γ (158). These observations can account for the higher risk of progression in HIV-1-infected individuals undergoing Th0/2-oriented immune responses. It is also of note that IL-4 up-regulates the expression by T cells of CD30 (147, 148), whose ligation induces enhanced HIV expression (149, 150) via NF-κB activation and HIV-1 long terminal repeat-driven transcription (149, 151). Thus, in HIV-1-infected individuals undergoing Th0/Th2-oriented immune responses at least two mechanisms that account for enhanced HIV-1 expression and progression towards full-blown AIDS can be operating. These findings can account for the results of several epidemiological studies showing that elevation in the serum of IgE antibody (which results from Th2–B cell collaboration) or heavy infestation by intestinal helminths (usually inducing striking Th2 responses) predict unfavourable prognosis in HIV-infected subjects (152–155). More importantly, they suggest that infections or vaccinations that preferentially stimulate Th2 cells, and perhaps even the heavy exposure to innocuous antigens (allergens) in atopic subjects, may increase the risk of progression in HIV-infected individuals and should therefore be avoided.

6. Concluding remarks

A large body of evidence has accumulated to suggest the existence of functionally polarized human Th cell responses based on their profile of cytokine secretion. Th1 cells produce IFN-γ, which activate macrophages, and are involved in phagocyte-dependent immune responses. By contrast, Th2 cells produce IL-4, and often also IL-5, IL-10, and IL-13, which are responsible for strong antibody responses and inhibit several macrophage functions. Th1 responses preferentially develop during infections by intracellular bacteria, whereas Th2 cells predominate during infestations by gastrointestinal nematodes. Several factors may influence the Th and Tc cell differentiation. They include the cytokine profile of 'natural immunity' evoked by different offending agents, the nature of the peptide ligand, as well as the activity of some co-stimulatory molecules and microenvironmentally secreted hormones, in the context of host genetic background.

In addition to playing different roles in protection, polarized Th1- and Th2-type responses can also be responsible for different types of immunopathological re-actions in humans. Th1-dominated responses are involved in the pathogenesis of multiple sclerosis, Type 1 diabetes mellitus, autoimmune thyroiditis, Crohn's disease, *H. pylori*-induced gastritis and peptic ulcer, acute kidney allograft rejection, un-explained recurrent abortions. In contrast, Th2-type responses predominate in trans-plantation tolerance, chronic graft-versus-host disease, progressive systemic sclerosis, and in HIV-infected patients showing more rapid evolution towards the full-blown disease. Moreover, allergen-specific Th2 responses are responsible for atopic dis-orders in genetically susceptible individuals.

The Th1/Th2 paradigm not only allows a better understanding of the main mechanisms involved in protection and of pathogenic mechanisms responsible for several immunopathic disorders, but also provides the basis for the development of new types of vaccines against infectious agents, as well as of novel strategies for the therapy of allergic and autoimmune disorders.

References

1. Mosmann, T. R., Cherwinski, H., Bond, and Coffman, R. L. (1986) Two types of murine T cell clones. Definition according to profiles of lymphokine activities and secreted proteins. *J. Immunol.*, **136**, 2348.
2. Abbas, A. K., Murphy, K. M., and Sher, A. (1996) Functional diversity of T lymphocytes. *Nature*, **383**, 787.
3. Romagnani, S. (1995) Biology of human Th1 and Th2 cells. *J. Clin. Immunol.*, **15**, 121.
4. Kelso, A. (1995) Th1 and Th2 subsets: paradigms lost? *Immunol. Today*, **16**, 374.
5. Hu-Li, J., Huang, H., Ryan, J., and Paul, W. E. (1997) In differentiated CD4$^+$ T cells, interleukin 4 production is cytokine-autonomous, whereas interferon gamma production is cytokine-dependent. *Proc. Natl. Acad. Sci. USA*, **94**, 3189.
6. Scheel, D., Richter, E., Toellner, K.-M., Reiling, N., Key, G., Wacker, H.-H., *et al.* (1995) Correlation of CD26 expression with Th1-like reactions in granulomatous diseases. In *Leukocyte typing V 'White cell differentiation antigens'* (ed. S. F. Schlossmann, L. Boumsell, W. Gilks, *et al.*), pp. 111–14. Oxford University Press, Oxford.
7. Annunziato, F., Manetti, R., and Tomasevic, L. (1996) Expression and release of LAG-3-associated protein by human CD4 T cells are associated with IFN-γ. *FASEB J.*, **10**, 767.
8. Bonecchi, R., Bianchi, G., Bordignon, P. P., *et al.* (1998) Differential expression of chemokines receptor and chemotactic responsiveness of type 1 T helper cells (Th1s) and Th2s. *J. Exp. Med.*, **187**, 129.
9. D'Elios, M. M., Romagnani, P., Scaletti, C., *et al.* (1997) *In vivo* CD30 expression in human diseases with predominant activation of Th2-like T cells. *J. Leuk. Biol.*, **61**, 539.
10. Sallusto, F., Mackay, C. R., and Lanzavecchia, A. (1997) Selective expression of the Eotaxin receptor CCR3 by human T helper 2 cells. *Science*, **277**, 2005.
11. Kamogawa, Y., Minasi, L. E., Carding, S. R., Bottomly, K., and Flavell, R. A. (1993) The relationship of IL-4- and IFN-γ-producing T cells studied by lineage ablation of IL-4-producing cells. *Cell*, **75**, 985.
12. Renz, H., Smith, H. R., Henson, J. E., Ray, B. S., Irvin, C. G., and Gelfand, E. W. (1992) Aerosolized antigen exposure without adjuvant causes increased IgE production and airways hyperresponsiveness in the mouse. *J. Allergy Clin. Immunol.*, **89**, 1127.

13. Xu-Amano, J., Kiyono, H., Jackson, R. J., Staats, H. F., Fujiashi, K., Burrows, P. D., *et al.* (1993) Helper T cell subsets for immunoglobulin A responses: oral immunization with tetanus toxoid and cholera toxin as adjuvant selectively induces Th2 cells in mucosa associated tissues. *J. Exp. Med.*, **178**, 1309.

14. Finkelman, F. D. (1995) Relationships among antigen presentation, cytokines, immune deviation, and autoimmune disease. *J. Exp. Med.*, **182**, 279.

15. Schweitzer, A. N., Borriello, F., Wong, R. C. K., Abbas, A. K., and Sharpe, A. H. (1997) Role of costimulators in T cell differentiation *J. Immunol.*, **158**, 2713.

16. Chace, J. H., Hooker, N. A., Mildenstein, K. L., Krieg, A. M., and Cowdery, J. S. (1997) Bacterial DNA-induced NK cell IFN-gamma production is dependent on macrophage secretion of IL-12. *Clin. Immunol. Immunopathol.*, **84**, 185.

17. Hosken, N. A., Shibuya, K., Heath, A. W., Murphy, K. M., and O'Garra, A. (1995) The effect of antigen dose on CD4$^+$ T helper cell phenotype development in a T cell receptor-αβ-transgenic model. *J. Exp. Med.*, **182**, 1579.

18. Pfeiffer, C., Murray, J., Madri, J., and Bottomly, K. (1991) Selective activation of Th1 and Th2 like cells *in vivo*. Response to human collagen IV. *Immunol. Rev.*, **123**, 65.

19. Tao, X., Constant, S., Jorristma, P., and Bottomly, K. (1997) Strength of TCR signal determines the costimulatory requirements for Th1 and Th2 CD4$^+$ T cell differentiation. *J. Immunol.*, **159**, 5956.

20. Manetti, R., Parronchi, P., Giudizi, M. G., Piccinni, M. P., Maggi, E., Trinchieri, G., *et al.* (1993) Natural killer cell stimulatory factor (interleukin 12) induces T helper Type 1 (Th1)-specific immune responses and inhibits the development of IL-4-producing Th cells. *J. Exp. Med.*, **177**, 1199.

21. Hsieh, C. S., Macatonia, S. E., Tripp, C. S., Wolf, S. F., O'Garra, A., and Murphy, K. M. (1993) Development of Th1 CD4$^+$ T cells through IL-12 produced by *Listeria*-induced macrophages. *Science*, **260**, 547.

22. Wenner, C., Guler, M. L., Macatonia, S. E., O'Garra, A., and Murphy, K. M. (1996) Role of IFN-γ and IFN-α in IL-12-induced Th1 development. *J. Immunol.*, **156**, 1442.

23. Parronchi, P., De Carli, M., Manetti, R., *et al.* (1992) IL-4 and IFN (α and γ) exert opposite regulatory effects on the development of cytolytic potential by Th1 or Th2 human T cell clones. *J. Immunol.*, **149**, 2977.

24. Rogge, L., Barberis-Maino, L., Biffi, M., Passini, N., Presky, D. H., Gubler, U., *et al.* (1997) Selective expression of an interleukin-12 receptor component by human T helper 1 cells. *J. Exp. Med.*, **185**, 825.

25. Yoshimoto, T., Okamura, H., Tagawa, Y. I., Iwakura, Y., and Nakanishi, K. (1997) Interleukin 18 together with IL-12 inhibits IgE production by induction of interferon-gamma production from activated B cells. *Proc. Natl. Acad. Sci. USA*, **15**, 3948.

26. Croft, M. and Swain, S. L. (1995) Recently activated naive CD4 T cells can help resting B cells, and can produce sufficient autocrine IL-4 to drive differentiation to secretion of T helper 2-type cytokines. *J. Immunol.*, **154**, 4269.

27. Seder, R. A., Germain, R. N., Lisley, P. S., and Paul, W. E. (1994) CD28-mediated costimulation of interleukin 2 (IL-2) production plays a critical role in T cell priming for IL-4 and interferon-γ production *J. Exp. Med.*, **179**, 299.

28. Kopf, M., Le Grost, G., Bachmann, M., Lamers, M. C., Bluethmann, H., and Kohler, G. (1993) Disruption of the murine IL-4 gene blocks Th2 cytokine responses. *Nature*, **362**, 245.

29. Maggi, E., Parronchi, P., Manetti, R., *et al.* (1992) Reciprocal regulatory effects of IFN-γ and IL-4 on the *in vitro* development of human Th1 and Th2 clones. *J. Immunol.*, **148**, 2142.

30. Conrad, D. H., Ben Sasson, S., Le Gros, G. G., Finkelman, F. D., and Paul, W. E. (1990)

Infection with *Nippostrongylus brasiliensis* or injection of anti-IgD antibodies markedly enhances Fc-receptor-mediated interleukin 4 production by non-B, non-T cells. *J. Exp. Med.*, **171**, 1497.

31. Aoki, I., Kinzer, C., Shirai, A., Paul, W. E., and Klinman, D. M. (1995) IgE receptor-positive non-B/non-T cells dominate the production of interleukin 4 and interleukin 6 in immunized mice. *Proc. Natl. Acad. Sci. USA*, **92**, 2534.

32. Bradding, P., Feather, I. H., Howarth, P. H., *et al.* (1992) Interleukin 4 is localized to and released by human mast cells. *J. Exp. Med.*, **176**, 1381.

33. Brunner, T., Heusser, C. H., and Dahinden, C. A. (1993) Human peripheral blood basophils primed by interleukin 3 (IL-3) produce IL-4 in response to immunoglobulin E receptor stimulation. *J. Exp. Med.*, **177**, 605.

34. Moqbel, R., Ying, S., Barkans, J., *et al.* (1995) Identification of messenger RNA for IL-4 in human eosinophils with granule localization and release of the translated product. *J. Immunol.*, **155**, 4939.

35. Werhil, B. K., Theodos, C. M., Galli, S. J., and Titus, R. G. (1994) Mast cells augment lesion size and persistence during experimental Leishmania major infection in the mouse. *J. Immunol.*, **152**, 4563.

36. Schmitz, J., Thiel, A., Kuhn, R., Rajewsky, K., Muller, W., Assenmacher, M., *et al.* (1994) Induction of interleukin 4 (IL-4) expression in T helper (Th) cells is not dependent on IL-4 from non-Th cells. *J. Exp. Med.*, **179**, 1349.

37. Bendelac, A., Killeen, N., Littman, D. R., and Schwartz, R. H. (1994) A subset of CD4[+] thymocytes selected by MHC class I molecules. *Science*, **263**, 1774.

38. Yoshimoto, T. and Paul, W. E. (1994) CD4[+], NK1.1[+] T cells promptly produce interleukin 4 in response to *in vivo* challenge with anti-CD3. *J. Exp. Med.*, **179**, 1285.

39. Yoshimoto, T., Bendelac, A., Watson, C., Hu-Li, J., and Paul, W. E. (1995) Role of NK1.1[+] T cells in a Th2 response and in immunoglobulin E production. *Science*, **270**, 1845.

40. Yoshimoto, T., Bendelac, A., Hu-Li, J., and Paul, W. E. (1995) Defective IgE production by SJL mice is linked to the absence of CD4[+], NK1.1[+] T cells that promptly produce interleukin 4. *Proc. Natl. Acad. Sci. USA*, **92**, 11931.

41. Launois, P., Ohteki, T., Swihart, K., Robson McDonald, H., and Louis, J. A. (1995) In susceptible mice, *Leishmania major* induce very rapid interleukin-4 production by CD4[+] T cells which are NK1.1-. *Eur. J. Immunol.*, **25**, 3298.

42. Pfeiffer, C., Stein, J., Southwood, S., Ketelaar, H., Sette, A., and Bottomly, K. (1995) Altered peptide ligands can control CD4 T lymphocyte differentiation *in vivo*. *J. Exp. Med.*, **181**, 1569.

43. Croft, M. and Swain, S. L. (1995) Recently activated naive CD4 T cells can help resting B cells, and can produce sufficient autocrine IL-4 to drive differentiation to secretion of T helper 2-type cytokines. *J. Immunol.*, **154**, 4269.

44. Kalinski, P., Hilkens, C. M. U., Wierenga, E. A., van der Pouw-Kraan, T. C. T. M., van Lier, R. A. W., Bos, J. D., *et al.* (1995) Functional maturation of human naive T helper cells in the absence of accessory cells. Generation of IL-4-producing T helper cells does not require exogenous IL-4. *J. Immunol.*, **154**, 3753.

45. Noben-Trauth, N., Shultz, L. D., Brombacher, F., Urban, J. F. Jr., Gu, H., and Paul, W. E. (1997) An interleukin 4 (IL-4)-independent pathway for CD4[+] T cell production is revealed in IL-4 receptor-deficient mice. *Proc. Natl. Acad. Sci. USA*, **94**, 10838.

46. Brikmann, V. and Kristofic, C. (1995) Regulation by corticosteroids of Th1 and Th2 cytokine production in human CD4[+] effector T cells generated from CD45RO– and CD45RO+ subsets. *J. Immunol.*, **155**, 3322.

47. Araneo, B. A., Dowell, T., Terui, T., Diegel, M., and Daynes, R. A. (1991) Dihydrotestosterone exerts a depressive influence on the production of IL-4, IL-5 and IFN-γ, but not IL-2 by activated murine cells. *Blood*, **78**, 688.

48. Daynes, R. A., Meikle, A. W., and Araneo, B. A. (1991) Locally active steroid hormones may facilitate compartmentalization of immunity by regulating the types of lymphokines produced by helper T cells. *Res. Immunol.*, **142**, 40.

49. Piccinni, M. P., Giudizi, M. G., Biagiotti, R., *et al.* (1995) Progesterone favours the development of human T helper (Th) cells producing Th2-type cytokines and promotes both IL-4 production and membrane CD30 expression in established Th1 clones. *J. Immunol.*, **155**, 128.

50. Piccinni, M.-P. and Romagnani, S. (1996) Regulation of fetal allograft survival by hormone-controlled Th1 and Th2-type cytokines. *Immunol. Res.*, **15**, 141.

51. Hsieh, C. S., Macatonia, S. E., O'Garra, A., and Murphy, K. M. (1995) T cell genetic background determines default T helper phenotype development *in vitro*. *J. Exp. Med.*, **181**, 713.

52. Prigent, P., Saoudi, A., Pannetier, C., Graber, G., Bonnefoy, J.-Y., and Druet, P. (1995) Mercuric chloride, a chemical responsible for T helper cell (Th2)-mediated autoimmunity in Brown Norways rats, directly triggers T cells to produce interleukin-4. *J. Clin. Invest.*, **96**, 1484.

53. Guler, M. L., Jacobson, N. G., Gluber, U., and Murphy, K. M. (1997) T cell genetic background determines maintenance of IL-12 signalling. *J. Immunol.*, **159**, 1767.

54. Takeda, K., Tanaka, T., Shi, W., *et al.* (1996) Essential role of Stat6 in IL-4 signalling. *Nature*, **380**, 627.

55. Shimoda, K., van Deursen, J., Sangster, M. Y., *et al.* (1996) Lack of IL-4-induced Th2 response and IgE class switching in mice with disrupted Stat6 gene. *Nature*, **380**, 630.

56. Li-Weber, M., Salgame, P., Hu, C., Davydov, I. V., and Krammer, P. H. (1997) Differential interaction of nuclear factor with the PRE-1 enhancer element of the human IL-4 promoter in different T cell subsets. *J. Immunol.*, **158**, 1195.

57. Hodge, M. R., Chun, H. J., Rengarajan, J., Alt, A., Liebsorn, R., and Glimcher, L. H. (1996) NF-AT-driven interleukin-4 transcription potentiated by NIP45. *Science*, **274**, 1903.

58. Ho, I. C., Hodge, M. R., Rooney, J. R., and Glimcher, L. H. (1996) The proto-oncogene c-*maf* is responsible for tissue-specific expression of interleukin-4. *Cell*, **85**, 973.

59. Ko, L. J., Yamamoto, M., Leonard, M. W., George, K. M., Ting, P., and Engel, J. D. (1991) Murine and human T-lymphocyte GATA-3 factors mediate transcription through a *cis*-regulatory element within the human T cell receptor d gene enhancer. *Mol. Cell. Biol.*, **11**, 2778.

60. Ting, C. N., Olson, M. C., Barton, K. P., and Leiden, J. M. (1996) Transcription factor GATA-3 is required for development of the T-cell lineage. *Nature*, **384**, 474.

61. Jacobson, N. G., Szabo, S. J., Weber-Nordt, R. M., Zhong, Z., Schreiber, R. D., Darnell, J. E., *et al.* (1995) Interleukin 12 signalling in T helper type 1 (Th1) cells involves tyrosine phosphorylation of signal transducer and activator of transcription (Stat)3 and Stat 4. *J. Exp. Med.*, **181**, 1755.

62. Szabo, S. J., Jacobson, N. G., Dighe, A. S., Gubler, U., and Murphy, K. M. (1995) Developmental commitment to the Th2 lineage by extinction of IL-12 signalling. *Immunity*, **2**, 666.

63. Bacon, C. M., McVicar, D. W., Ortaldo, J. R., Rees, R. C., O'Shea, J. J., and Johnston, J. A. (1995) Interleukin 12 (IL-12) induces tyrosine phosphorylation of JAK2 and TYK2: differential use of Janus family tyrosine kinases by IL-2 and IL-12. *J. Exp. Med.*, **181**, 399.

64. Takeda, K., Tanaka, T., Shi, W., *et al.* (1996) Essential role of Stat6 in IL-4 signalling. *Nature*, **380**, 627.

65. Seder, R. A. (1995) The role of IL-12 in the regulation of Th 1 and Th2 differentiation. *Res. Immunol.*, **146**, 473.

66. Fiorentino, D. F., Zlotnik, A., Vieira, P., Mosmann, T. R., Howard, M., Moore, K. W., *et al.* (1991) IL-10 acts on the antigen-presenting cell to inhibit cytokine production by Th1 cells. *J. Immunol.*, **146**, 3444.

67. Trinchieri, G. and Scott, P. (1995) Interleukin-12: a pro-inflammatory cytokine with immunoregulatory functions. *Res. Immunol.*, **146**, 423.

68. Seder, R. A., Boulay, J. L., Finkelman, F., Babier, S., Ben-Sasson, S. Z., Le Gros, G. G., *et al.* (1992) CD8$^+$ T cells can be primed *in vitro* to produce IL-4. *J. Immunol.*, **148**, 1652.

69. Gajewski, T. F. and Fitch, F. W. (1990) Anti-proliferative effect of IFN-gamma in immune regulation. IV. Murine CTL clones produce IL-3 and GM-CSF, the activity of which is masked by the inhibitory action of secreted IFN-gamma. *J. Immunol.*, **144**, 548.

70. Abed, N. S., Chace, J. H., and Cowdery, J. S. (1994) T cell-independent and T cell-dependent B cell activation increases IFN-gamma receptor expression and renders B cells sensitive to IFN-gamma-mediated inhibition. *J. Immunol.*, **153**, 3369.

71. Finkelman, F. D., Svetic, A., Gresser, I., Snapper, C., Holmes, J., Trotta, P. P., *et al.* (1991) Regulation by interferon of immunoglobulin isotype selection and lymphokine production in mice. *J. Exp. Med.*, **174**, 1179.

72. Parronchi, P., Mohapatra, S., Sampognaro, S., *et al.* (1996) Effects of interferon-α repertoire and peptide reactivity of human allergen-specific T cells. *Eur. J. Immunol.*, **26**, 697.

73. Seder, R. A., Gazzinelli, R., Sher, A., and Paul, W. E. (1993) Interleukin 12 acts directly on CD4$^+$ T cells to enhance priming for interferon g production and diminishes interleukin 4 inhibition of such priming. *Proc. Natl. Acad. Sci. USA*, **90**, 10188.

74. Wynn, T. A., Jankovic, D., Hieny, S., Zioncheck, K., Jardieu, P., Cheever, A. W., *et al.* (1995) IL-12 exacerbates rather than suppresses T helper 2-dependent pathology in the absence of endogenous IFN-γ. *J. Immunol.*, **154**, 3999.

75. Nickerson, P., Steurer, W., Steiger, J., Zheng, X., Steele, A. W., and Strom, T. B. (1994) Cytokines and the Th1/Th2 paradigm in transplantation. *Curr. Opin. Immunol.*, **6**, 757.

76. Suthanthiran, M. and Strom, T. B. (1995) Immunobiology and immunopharmacology of organ allograft rejection. *J. Clin. Immunol.*, **15**, 161.

77. D'Elios, M. M., Josien, R., Manghetti, M., *et al.* (1997) Predominant Th1 cell infiltration in acute rejection episodes of human kidney grafts. *Kidney Int.*, **51**, 1876.

78. Lin, H., Mosmann, T. R., Guilbert, L., Tuntipopipat, S., and Wegmann, T. G. (1993) Synthesis of T helper 2-type cytokines at maternal–fetal interface. *J. Immunol.*, **151**, 4562.

79. Wegmann, T. G., Lin, H., Guilbert, L., and Mossmann, T. R. (1993) Bidirectional cytokine interactions in the maternal–fetal relationship: is successful pregnancy a Th2 phenomenon? *Immunol. Today*, **14**, 353.

80. Romagnani, S. and Maggi, E. (1994) Biology of human Th1 and Th2 responses in AIDS. *Curr. Opin. Immunol.*, **6**, 616.

81. Wierenga, E. A., Snoek, M., De Groot, C., Chrètien, I., Bos, J. D., Jansen, H. M., *et al.* (1990) Evidence for compartmentalization of functional subsets of CD4$^+$ T lymphocytes in atopic patients. *J. Immunol.*, **144**, 4651.

82. Parronchi, P., Macchia, D., Piccinni, M. P., Biswas, P., Simonelli, C., Maggi, E., *et al.* (1991) Allergen- and bacterial antigen-specific T-cell clones established from atopic donors show a different profile of cytokine production. *Proc. Natl. Acad. Sci. USA*, **88**, 4538.

83. Maggi, E., Biswas, P., Del Prete, G. F., *et al.* (1991) Accumulation of Th2-like helper T cells in the conjunctiva of patients with vernal conjunctivitis. *J. Immunol.*, **146**, 1169.

84. Varney, V. A., Jacobson, R., Sudderinck, M. R., Robinson, D. S., Irani, A.-M. A., Scwartz, L. B., *et al.* (1992) Immunohistology of the nasal mucosa following allergen-induced rhinitis: identification of activated T lymphocytes, eosinophils and neutrophils. *Am. Rev. Respir. Dis.*, **146**, 170.

85. Del Prete, G. F., De Carli, M., D'Elios, M. M., Maestrelli, P., Ricci, M., Fabbri, L., *et al.* (1993) Allergen exposure induced the activation of allergen-specific Th2 cells in the airway mucosa of patients with allergenic respiratory disorders. *Eur. J. Immunol.*, **23**, 1445.

86. Kay, A. B., Ying, S., Varney, V., Gaga, M., Durham, S. R., Moqbel, R., *et al.* (1991) Messenger RNA expression of the cytokine gene cluster, interleukin 3 (IL-3), IL-4, IL-5 and granulocyte/macrophage colony-stimulating-factor, in allergen-induced late-phase reactions in atopic subjects. *J. Exp. Med.*, **173**, 775.

87. Hamid, Q., Azzawi, M., Ying, S., *et al.* (1991) Expression of mRNA for interleukin-5 in mucosal bronchial biopsies from asthma. *J. Clin. Invest.*, **87**, 1541.

88. Robinson, D., Hamid, Q., Ying, S., Tsicopoulos, A., Barkans, J., Bentley, A., *et al.* (1992) Predominant T_{H2}-like bronchoalveolar T-lymphocyte population in atopic asthma. *N. Engl. J. Med.*, **326**, 298.

89. Robinson, D., Hamid, Q., Ying, S., Bentley, A., Assoufi, B., Durham, S. R., *et al.* (1993) Prednisolone treatment in asthma is associated with modulation of bronchoalveolar lavage cell interleukin-4, interleukin-5, and interferon-γ cytokine gene expression. (1993) *Am. Rev. Respir. Dis.*, **148**, 401.

90. Ying, S., Durham, S. R., Corrigna, C. J., Hamid, Q., and Kay, A. B. (1995) Phenotype of cells expressing mRNA for Th2-type (interleukin-4 and interleukin-5) and Th1-type (interleukin-2 and interferon-γ) cytokines in bronchoalveolar lavage and bronchial biopsies from atopic asthmatics and normal control subjects. *Am. J. Respir. Cell. Biol.*, **12**, 477.

91. Tang, M. and Kemp, A. (1994) Production and secretion of interferon-gamma in children with atopic dermatitis. *Clin. Exp. Immunol.*, **95**, 66.

92. Kimata, H., Fujimoto, M., and Furusho, K. (1995) Involvement of interleukin (IL)-13, but not IL-4, in spontaneous IgE and IgG4 production in nephrotic syndrome. *Eur. J. Immunol.*, **25**, 1497.

93. Van Reijsen, F. C., Bruijnzeel-Koomen, C. A. F. M., Kalthoff, F. S., Maggi, E., Romagnani, S., Westland, J. K. T., *et al.* (1992) Skin-derived aeroallergen-specific T-cell clones of Th2 phenotype in patients with atopic dermatitis. *J. Am. Clin. Immunol.*, **90**, 184.

94. Reinhold, U., Kukel, S., Goeden, B., Neumann, C., and Kreysel, H. W. (1991) Functional characterization of skin-infiltrating lymphocytes in atopic dermatitis. *Clin. Exp. Immunol.*, **86**, 444.

95. Sager, N., Feldmann, A., Schilling, G., Kreitsch, P., and Neumann, C. (1992) House dust mite-specific T cells in the skin of subjects with atopic dermatitis: frequency and lymphokine profile in the allergen patch test. *J. Allergy Clin. Immunol.*, **89**, 801.

96. Gutgesell, C., Yssel, H., Scheel, D., Gerdes, J., and Neumann, C. (1994) IL-10 secretion of allergen-specific skin-derived T cells correlates positively with that of the Th2 cytokines IL-4 and IL-5. *Exp. Dermatol.*, **3**, 304.

97. Virtanen, T., Maggi, E., Manetti, R., Piccinni, M.-P., Sampognaro, S., Parronchi, P., *et al.* (1995) No relationship between skin-infiltrating Th2-like cells and allergen-specific IgE response in atopic dermatitis. *J. Allergy Clin. Immunol.*, **96**, 411.

98. Thepen, T., Langeveld-Wildschut, E. G., Bihari, I. C., van Wichen, D. F., van Reijsen, F. C.,

Mudde, G. C., *et al.* (1996) Biphasic response against aeroallergen in atopic dermatitis showing a switch from an initial Th2 response into Th1 response *in situ*. An immuno-cytochemical study. *J. Am. Clin. Immunol.*, **97**, 828.

99. Romagnani, S. (1997) *Molecular biology intelligence unit. The Th1/Th2 paradigm in disease.* Autoimmune disorders, pp. 131–45. Springer. R. G. Landes Company, Austin, Texas.

100. Del Prete, G. F., Tiri, A., Mariotti, S., Pinchera, A., Ricci, M., and Romagnani, S. (1987) Enhanced production of gamma-interferon by thyroid-derived T cell clones from patients with Hashimoto's thyroiditis. *Clin. Exp. Immunol.*, **69**, 323.

101. Del Prete, G. F., Tiri, A., De Carli, M., Mariotti, S., Pinchera, A., Chretien, I., *et al.* (1989) High potential to tumor necrosis factor a (TNF-α) production of thyroid infiltrating lymphocytes in Hashimoto's thyroiditis: a peculiar feature of destructive thyroid autoimmunity. *Autoimmunity*, **4**, 267.

102. Zheng, R. Q. H., Abney, E. R., Chu, C. Q., Field, M., Maini, R. N., Lamb, J. R., *et al.* (1992) Detection of *in vivo* production of tumor necrosis factor-alpha by human thyroid epithelial cells. *Immunology*, **75**, 456.

103. De Carli, M., D'Elios, M. M., Mariotti, S., Marcoscci, C., Pinchera, A., Ricci, M., *et al.* (1993) Cytolytic T cells with Th 1-like cytokine profile predominate in retroorbital Iymphocytic infiltrates of Graves' ophthalmopathy. *J. Clin. Endocrinol. Metab.*, **77**, 1120.

104. Watson, P. F., Pickering, P., Davies, R., and Weetman, A. P. (1994) Analysis of cytokine gene expression in Graves' disease and multinodular goiter. *J. Clin. Endocrinol. Metab.*, **79**, 355.

105. Paschke, R., Schuppert, E., Taton, M., and Velu, T. (1994) Intrathyroidal cytokine gene expression profiles in autoimmune thyroiditis. *J. Endocrinol.*, **41**, 309.

106. McLachlan, S. M., Prummel, M. F., and Rapoport, B. (1994) Cell-mediated or humoral immunity in Graves' ophthalmopathy? Profiles of T-cell cytokines amplified by polymerase chain reaction from orbital tissue. *J. Clin. Endocrinol. Metab.*, **78**, 1070.

107. Grubeck-Loebenstein, B., Trieb, K., Sztankay, A., Holter, W., Anderl, H., and Wick, G. (1994) Retrobulbar T cells from patients with Graves' ophthalmopathy are CD8[+] and specifically recognize autologous fibroblasts. *J. Clin. Invest.*, **93**, 2738.

108. Roura-Mir, C., Catalfamo, M., Lucas-Martin, A., Sospedra, M., Pujol-Borrell, R., and Jaraquemada, D. (1996) IL-4/IFN-γ balance in Graves' disease contrasts with IFN-γ dominance in Hashimoto's thyroiditis. *Proc. IV Int. Conf. on Cytokines*. Florence, March 20–22. (Abstract).

109. Mullins, R. J., Cohen, S. B. A., Webb, L. M. C., Chernajovsky, Y., Dayan, C. M., Londei, M., *et al.* (1995) Identification of thyroid stimulating hormone receptor-specific T cells in Graves' disease thyroid using autoantigen-transfected Epstein Barr virus-transformed B cell lines. *J. Clin. Invest.*, **96**, 30.

110. Chofflon, M., Juillard, C., Juillard, P., Gauthier, G., and Grau, G. E. (1992) Tumor necrosis factor alpha production as a possible predictor of relapse in patients with multiple sclerosis. *Eur. Cytokine Netw.*, **3**, 523.

111. Rieckmann, P., Albrecht, M., Kitze, B., Weber, T., Tumani, H., Broocks, A., *et al.* (1994) Cytokine mRNA levels in mononuclear blood cells from patients with multiple sclerosis. *Neurology*, **44**, 1523.

112. Rieckmann, P., Albrecht, M., Kitze, B., *et al.* (1995) Tumor necrosis factor alpha messenger RNA expression in patients with relapsing remitting multiple sclerosis is associated with disease activity. *Ann. Neurol.*, **37**, 82.

113. Brod, S., Benjamin, D., and Hafler, D. A. (1991) Restricted T cell expression of IL2/IFN-γ mRNA in human inflammatory disease. *J. Immunol.*, **147**, 810.

114. Benvenuto, R., Paroli, M., Buttinelli, C., Franco, A., Barnaba, V., Fieschi, C., *et al.* (1991) Tumor necrosis factor-alpha synthesis by cerebrospinal-sclerosis. Fluid-derived T cell clones from patients with multiple. *Clin. Exp. Immunol.*, **84**, 97.

115. Olsson, T. (1993) Cytokines in neuroinflammatory disease: role of myelin- autoreactive T cell production of interferon-gamma. *J. Neuroimmunol.*, **40**, 211.

116. Correale, J., Gilmore, W., McMillan, M., Li, S., McCarthy, K., Le, T., *et al.* (1995) Patterns of cytokine secretion by autoreactive proteolipid protein-specific T cell clones during the course of multiple sclerosis. *J. Immunol.*, **154**, 2959.

117. Selmaj, K., Raine, C. S., Cannella, B., and Brosnan, C. F. (1991) Identification of lymphotoxin and tumor necrosis factor in multiple sclerosis lesions. *J. Clin. Invest.*, **87**, 949.

118. Windhagen, A., Newcombe, J., Dangond, F., Strand, C., Woodroofe, M. N., Cuzner, M. L., *et al.* (1995) Expression of costimulatory molecules B7–1 (CD80), B7–2 (CD86), and interleukin-12 cytokine in multiple sclerosis lesions. *J. Exp. Med.*, **182**, 1985.

119. Panitch, H. S., Hirsch, R. L., Schindler, J., and Johnson, K. P. (1987) Treatment of multiple sclerosis with gamma-interferon: exacerbations associated with activation of the immune system. *Neurology*, **37**, 1097.

120. The IFN B Multiple Sclerosis Study Group. (1993) Interferon beta-1 β is effective in relapsing-remitting multiple sclerosis. I. Clinical results of a multicenter, randomized, double-blind, placebo-controlled trial. *Neurology*, **43**, 655.

121. Zipris, D., Greiner, D. L., Malkani, S., Whalen, B., Mordes, J. P., and Rossini, A. A. (1996) Cytokine gene expression in islets and thyroids of BB rats. IFN-γ and IL-1 2p40 mRNA increase with age in both diabetic and insulin-treated nondiabetic BB rats. *J. Immunol.*, **156**, 1315.

122. Wong, F. S., Visintin, I., Wen, L., Flavell, R. A., and Janeway, C. A. Jr. (1996) CD8 T cell clones from young nonobese diabetic (NOD) islets can transfer rapid onset of diabetes in NOD mice in the absence of CD4 cells. *J. Exp. Med.*, **183**, 67.

123. O'Hara, R. M. and Henderson, S. L. (1995) *FASEB J.*, **9**, 5938 (Abstract).

124. Foulis, A. K., McGill, M., and Farquharson, M. A. (1991) Insulitis in type 1 (insulin-dependent) diabetes mellitus in man. Macrophages, lymphocytes, and interferon-gamma containing cells. *J. Pathol.*, **165**, 97.

125. Chang, J. C., Linarelli, L. G., Laxer, J. A., Froning, K. J., Caralli, L. L., Brostoff, S. W., *et al.* (1995) Insulin-secretory-granule specific T cell clones in human insulin dependent diabetes mellitus. *J. Autoimm.*, **8**, 221.

126. Postlethwhaite, A. E. and Seyer, J. M. (1991) Fibroblast chemotaxis induction by human recombinant interleukin 4: identification by synthetic peptide analysis of two chemotactic domains residing in amino acid sequences 70–88 and 89–122. *J. Clin. Invest.*, **87**, 2147.

127. Postlethwhaite, A. E., Holness, A., Katai, H., and Raghow, R. (1992) Human fibroblasts synthesize elevated levels of extracellular matrix proteins in response to interleukin 4. *J. Clin. Invest.*, **90**, 1479.

128. Gillery, P., Fertin, C., Nicolas, J. F., Chastang, F., Kalis, B., Banchereau, J., *et al.* (1992) Interleukin-4 stimulates collagen gene expression in human fibroblast monolayer cultures. *Fed. Eur. Biochem. Sci.*, **302**, 231.

129. Famularo, G., Procopio, A., Giacomelli, R., Danese, C., Sacchetti, S., Perego, M. A., *et al.* (1990) Soluble interleukin-2 receptor, interleukin-2 and interleukin-4 in sera and supernatants from patients with progressive systemic sclerosis. *Clin. Exp. Immunol.*, **81**, 368.

130. Mavilia, C., Scaletti, C., Romagnani, P., *et al.* (1997) Type 2 helper T-cell predominance and high CD30 expression in Systemic sclerosis. *Am. J. Pathol.*, **151**, 1751.

131. D'Elios, M. M., Manghetti, M., De Carli, M., Costa, F., Baldari, C. T., Burroni, D., *et al.* (1997) T helper 1 effector cells specific for *Helicobacter pylori* in the gastric antrum of patients with peptic disease. *J. Immunol.*, **158**, 962.

132. Sadlack, B., Merz, H., Schorle, H., Scimpl, A., Feller, A. C., and Horvak, I. (1993) Ulcerative colitis-like disease in mice with a disrupted interleukin-2 gene. *Cell*, **75**, 253.

133. Kuhn, R., Rajewski, K., and Muller, W. (1991) Generation and analysis of interleukin-4 deficient mice. *Science*, **254**, 707.

134. Kuhn, R., Lohler, J., Rennick, D., Rajewski, K., and Muller, W. (1993) Interleukin-10-deficient mice develop chronic enterocolitis. *Cell*, **75**, 263.

135. Kulkarni, A. B., Huh, C. G., Becker, D., *et al.* (1993) Transforming growth factor β1 null mutation in mice causes excessive inflammatory response and early death. *Proc. Natl. Acad. Sci. USA*, **90**, 770.

136. Rudolph, U., Finegold, M. J., Rich, S. S., Harriman, G. R., Srinivasan, Y., Brabet, P., *et al.* (1995) Ulcerative colitis and adenocarcinoma of the colon in G alpha i2-deficient mice. *Nature Genet.*, **10**, 143.

137. Neurath, M. F., Fuss, I., Kelsall, B. L., Stuber, E., and Strober, W. (1995) Antibodies to interleukin 12 abrogate established experimental colitis in mice. *J. Exp. Med.*, **182**, 1281.

138. Powrie, F., Correa-Oliveira, R., Mauze, S., and Coffman, R. L. (1994) Regulatory interactions between CD45RBhi and CD45RBlo CD4$^+$ T cells are important for the balance between protective and pathogenic cell-mediated immunity *J. Exp. Med.*, **179**, 589.

139. Romagnani, S. (1995) Atopic allergy and other hypersensitivities. *Curr. Opin. Immunol.*, **7**, 745.

140. Parronchi, P., Romagnani, P., Annunziato, F., *et al.* (1997) Type 1 T-helper cell predominance and interleukin-12 expression in the gut of patients with Crohn's disease. *Am. J. Pathol.*, **150**, 823.

141. Clerici, M. and Shearer, G. M. (1993) Th1/Th2 switch is a critical step in the aetiology of HIV infection. *Immunol. Today*, **14**, 107.

142. Graziosi, C., Pantaleo, G., Gantt, K. R., Fortin, J.-P., Demarest, J. F., Cohen, O. J., *et al.* (1994) Lack of evidence for the dichotomy of Th1 and Th2 predominance in HIV-infected individuals. *Science*, **265**, 248.

143. Maggi, E., Mazzetti, M., Ravina, A., *et al.* (1994) Ability of HIV to promote a Th1 to Th0 shift and to replicate preferentially in Th2 and Th0 cells. *Science*, **265**, 244.

144. Vyakarnam, A., Matear, P. M., Martin, S. J., and Wagstaff, M. (1995) Th1 cells specific for HIV-1 gag p24 are less efficient than Th0 cells in supporting HIV replication, and inhibit virus replication in Th0 cells. *Immunology*, **86**, 85.

145. Tanaka, Y., Koyanagi, Y., Tanaka, R., Kumazawa, Y., Nishimura, T., and Yamamoto, N. (1997) Productive and lytic infection of human CD4$^+$ type helper T cells with macrophage-tropic human immunodeficiency virus type 1. *J. Virol.*, **71**, 465.

146. Choe, H., Farzan, M., Sun, Y., *et al.* (1996) The beta-chemokine receptors CCR3 and CCR5 facilitate infection by primary HIV-1 isolates. *Cell*, **85**, 1135.

147. Nakamura, T., Lee, N. K., Nam, S. Y., Al-Ramadi, B. K., Koni, P. A., Bottomly, K., *et al.* (1997) Reciprocal regulation of CD30 expression on CD4$^+$ T cells by IL-4 and IFN-gamma. *J. Immunol.*, **158**, 2090.

148. Maggi, E., Manetti, R., Annunziato, F., Cosmi, L., Giudizi, M.-G., Biagiotti, R., *et al.* (1997) Functional characterization and modulation of cytokine production by CD8$^+$ T cells from human immunodeficiency virus-infected individuals. *Blood*, **89**, 3672.

149. Biswas, P., Smith, C. A., Goletti, D., Hardy, E. C., Jackson, R. W., and Fauci, A. S. (1995) Cross-linking of CD30 induces HIV expression in chronically infected T cells. *Immunity*, **2**, 587.

150. Maggi, E., Annunziato, F., Manetti, R., Biagiotti, R., Giudizi, M.-G., Ravina, A., *et al.* (1995) Activation of HIV expression by CD30 triggering in CD4$^+$ cells from HlV-infected individuals. *Immunity*, **3**, 251.

151. McDonald, P. P., Cassatella, M. A., Bald, A., Maggi, E., Romagnani, S., Gruss, H. J., *et al.* (1995) CD30 ligation induces nuclear factor-κB activation in human T cell lines *J. Immunol.*, **25**, 2870.

152. Romagnani, S., Maggi, E., Del Prete, G. F., *et al.* (1994) Role of Th1/Th2 cytokines in HIV infection. *Immunol. Rev.*, **140**, 73.

153. Israel-Biet, D., Labrousse, F., Tourani, J.-M., Sors, H., Andrieu, J.-M., and Even, P. (1992) Elevation of IgE in HlV-infected subjects: a marker of poor prognosis. *J. Allergy Clin. Immunol.*, **89**, 68.

154. Viganò, A., Principi, N., Crupi, L., Onorato, J., Vincenzo, Z. G., and Salvaggio, A. (1995) Elevation of IgE in HIV-infected children and its correlation with the progression of disease. *J. Allergy Clin. Immunol.*, **95**, 627.

155. Bentwich, Z., Kalinkovic, A., and Weisman, Z. (1995) Immune activation is a dominant factor in the pathogenesis of African AIDS. *Immunol. Today*, **16**, 187.

156. Piccinni, M.-P., Beloni, L., Livi, C., Maggi, E., Scarselli, G., Romagnani, S. (1998). Role of type 2 T helper (Th2) cytokines and leukemia inhibitory factor (LIF) produced by decidual T cells in the development and maintenance of successful pregnancy. *Nature Med.* **4**, 1020.

157. Mombaerts, P., Mizeguelin, E., Guisby, M. J., Glimcher, L. H., Bahn, A. K., and Tonegauwe, S. (1993) Spontaneous development of inflammatory back disease in T cell receptor mutant mice, **75**, 275.

158. Galli, G., Annunziato, F., Mavilia, C., Romagnani, P., Cosmi, L., Manetti, R., Pupilli, C., Maggi, E., and Romagnani, S. (1998) Enhanced HIV expression during T helper 2-oriented responses due to the opposite regulatory effect of IL-4 and IFN-γ on Fusin/CXCR4. *Eur. J. Immunol.* **28**, 3280.

5 | Chemokines

ALBERTO MANTOVANI AND SILVANO SOZZANI

1. Introduction

Leukocyte recruitment has been recognized as an early event in inflammatory processes since the late 19th century. Accumulation and trafficking of leukocytes in tissues under physiological and pathological conditions are orderly (typically neutrophils precede mononuclear cells) and selective because in certain states one or more leukocyte subset is recruited preferentially (e.g. eosinophils in allergy). The current paradigm of recruitment is that of a multistep process involving the action of chemotactic signals (1, 2).

Classical chemoattractants include complement components, formyl peptides, and leukotriene B4. In addition, various cytokines are able to elicit directional migration of leukocytes. Whereas molecules, such as monocyte-colony stimulating factor or tumour necrosis factor, exert chemotactic activity; the main chemotactic cytokines are a superfamily of molecules known as chemokines (for chemotactic cytokines).

Chemokines play a central role in the multistep process of leukocyte recruitment and are involved in a variety of disease processes including inflammation, allergy, and neoplasia. As such they now represent a prime target for the development of novel therapeutic strategies. Chemical antagonists of chemokines remain the holy grail for the future.

Several independent lines of work lead to the identification of chemokines such as monocyte chemotactic protein-1 (MCP-1) and related molecules. Already in the early 1970s it had been noted that supernatants of activated blood mononuclear cells contained attractants active on monocytes and neutrophils (3). Subsequently, a chemotactic factor active on monocytes was identified in culture supernatants of mouse (4) and human (5, 6) tumour lines, and was called tumour-derived chemotactic factor (TDCF) (5–7). TDCF was at the time rather unique in that it was active on monocytes but not on neutrophils (6) and had a low molecular weight (5, 6). Moreover, correlative evidence suggested its involvement in the regulation of macrophage infiltration in murine and human tumours (5, 6, 8). A molecule with similar cellular specificity and physicochemical properties was independently identified in the culture supernatant of smooth muscle cells (SMDCF) (9). The JE gene had been identified as an immediate-early PDGF-inducible gene in fibroblasts (10, 11). Thus, in the mid-1980s a gene (JE) was in search of function, and a monocyte-specific attractant was waiting for molecular definition. In 1989, MCP-1 was successfully purified from supernatants of a human glioma (12) a human monocytic leukaemia (13) and a human sarcoma cell line (14–16): sequencing and molecular cloning revealed its relationship with the long known JE gene (17–19).

A comprehensive review of the chemokine world is beyond the scope of this chapter, given the wealth of information accumulated in recent years. Here we will concisely summarize the main general features of this 'world apart' and focus on selected aspects (e.g. role in polarized Th1 versus Th2 responses; trafficking of dendritic cells) that have attracted our attention. For more extensive coverage of the literature we refer the reader to recent reviews (1–3, 20–24).

2. Structure and families

Chemokines are a superfamily of small proteins that play a crucial role in immune and inflammatory reactions and in viral infection (21–23). Most chemokines cause chemotactic migration of leukocytes, but these molecules also affect angiogenesis, collagen production, and the proliferation of haematopoietic precursors. Based on a cysteine motif, a CXC, CC, C, and CX3C family have been identified (Fig. 1). The chemokine scaffold consists of an N-terminal loop connected via Cys bonds to the more structured core of the molecule (three β-sheets) with a C-terminal α-helix. About 50 human chemokines have been identified.

The major eponymous function of chemokines is chemotaxis for leukocytes. Schematically (Fig. 1), CXC (or α) chemokines are active on neutrophils (PMN) and lymphocytes whereas CC (or β) chemokines exert their action on multiple leukocyte

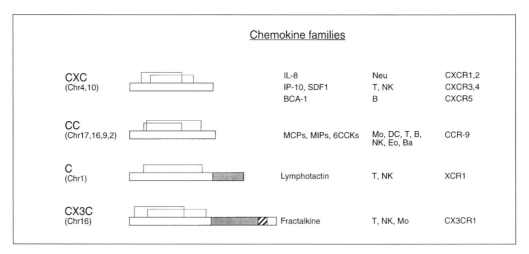

Fig. 1 A schematic, simplified view of chemokine families. Acronyms and chromosomes refer to man. Only prototypic molecules or family of molecules and main cellular targets are shown. Clusters of related molecules are identified as monocyte chemotactic proteins (MCPs, 1–4), macrophage inflammatory proteins (MIPs), molecules with a six Cys motif (6 CCK). Certain molecules are bigger (lymphotactin, fractalkine), with a C-terminal, mucin-like domain (dotted part). Fractalkine is expressed as a membrane molecule, with a transmembrane domain, on activated endothelial cells. The chromosomal localization is shown in parenthesis and receptors are in the right-hand column.

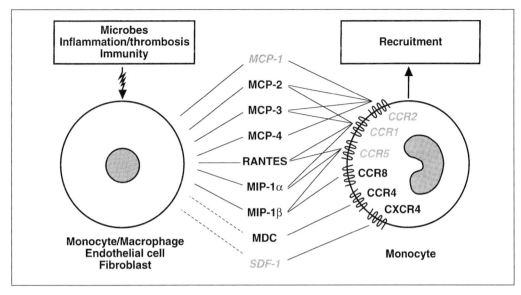

Fig. 2 Redundancy of chemokines in monocyte recruitment. Resting or stimulated mononuclear phagocytes and stromal cells (endothelial cells, fibroblasts) concomitantly produce many cytokines (polyspeirism) whose spectrum of action includes monocytes. Production is constitutive (dotted line) or inducible (continuous line). Monocytes express several chemokine receptors that are recognized in a promiscuous way. The italic indicates that knock-out mice or deficient humans (Δ 32 *CCR5*) have been studied. The system is built so that inactivation of any single component (agonist or receptor) allows a minimal level of phagocyte trafficking sufficient for fundamental functions, such as tissue remodelling, clearance of debris, and innate resistance.

subtypes, including monocytes, basophils, eosinophils, T lymphocytes, dendritic cells, and NK cells, but they are generally inactive on PMN. Eotaxins (CC) represent the chemokines with the most restricted spectrum of action being selectively active on eosinophilic and basophilic granulocytes (20, 21, 24, 25). Lymphotactin and fractalkine are the only proteins so far described with a C and CX3C motif, respectively (26–28). They both act on lymphoid cells (T lymphocytes and NK cells) and fractalkine is also active on monocytes and NK cells (26–30) (and Sozzani, Allavena and Wells, unpublished).

Chemokines are redundant in their action on target cells (see for instance for monocytes, Fig. 2). No chemokine is active only on one leukocyte population and usually a given leukocyte population has receptors for and responds to different molecules (Table 1). Interestingly, mononuclear phagocytes, the most evolutionary ancient cell type of innate immunity, respond to the widest range of chemokines. These include most CC chemokines, fractalkine (CX3C), and certain CXC molecules (e.g. SDF1 and, under certain conditions, IL-8).

3. Receptors and signal transduction

All chemokine receptors identified to date are seven transmembrane domain proteins coupled to GTP-binding proteins with homology to the family of chemotactic re-

Table 1 Expression of chemokine receptors in human leukocyte populations

Receptor	Main ligands	Main cells[a]
CCR1	MCP-3, RANTES, MIP-1α	Mo, T, NK, DC, PMN
CCR2 B/A	MCP-s (1–4)	Mo, T (act.), NK (act.)
CCR3	Eotaxin, MCP-3, RANTES	Eo, Ba, T (Th2)
CCR4	TARC, MDC	T (Th2, Tc2), NK, DC
CCR5	MIP-1β, MIP-1α, RANTES	Mo, T (Th1, Tc1), DC
CCR6	MIP-3α/LARC/Exodus	T, DC (CD34)
CCR7	ELC, SLC	T, Mo
CCR8	1309, TARC	T (Th2), Mo
CCR9	TECK	T
CXCR1	IL-8, GCP-2	PMN
CXCR2	IL-8, gro, NAP-2, GCP-2	PMN
CXCR3	IP10, MIG	T (Th1)
CXCR4	SDF-1	Widely expressed
CXCR5	BCA-1	B
CXCR1	Lymphotactin	T, NK
CX3CR1	Fractakline	Mo, NK, T

[a] Mo, monocytes; DC, dendritic cells; DC (CD34), DC derived from CD34 cells *in vitro*; PMN, neutrophils; Eo, eosinophils; Ba, basophils; Th, T helper; Tc, T cytotoxic; (act.), activated.

ceptors. Five receptors for CXC chemokines (CXCR1–5) and nine for CC chemokines (CCR1–9) were recently cloned (Table 1). These receptors show a promiscuous pattern of ligand recognition and are differentially expressed and regulated in leukocytes (21, 23, 31–33). The receptor for fractalkine and lymphotactin have been recently characterized (30, 34).

Activation of chemotactic receptors results in an increase of intracellular free calcium concentration and in the activation of multiple phospholipases (C, D, and A2). Calcium fluxes were observed after receptor binding by most of the CC chemokines (20, 35–38). This response is rapid, transient, and sensitive to *Bordetella pertussis* toxin (PTox) (39–43), which suggest that chemokine receptors are associated with PTox-sensitive GTP-binding proteins. In support of this observation it was reported that monocyte chemotaxis in response to MCP-1, MCP-3, RANTES, and MIP-1α is inhibited in a concentration-dependent manner by PTox (39, 44, 45), whereas, in the same experimental conditions, *Cholera* toxin was ineffective (39, 45). Studies performed in cell transfectants have shown that both CXCR1 and CXCR2, as well as CCR1 and CCR2, can efficiently couple with PTox-insensitive G-protein, such as Gα14 and Gα16 (46, 47). The biological relevance of this finding for circulating leukocytes is still unclear.

In human monocytes activated by MCP-1, the influx of Ca^{2+} across plasma membrane, rather than the release from intracellular stores, appeared to be the main mechanism responsible for intracellular Ca^{2+} elevation (44, 45). Calcium influx was required for arachidonate accumulation by MCP-1 in human monocytes (48). Bioactive products of arachidonic acid metabolism (PAF and 5-oxo-ETE) increased, in a synergistic fashion, both arachidonic acid release and chemotactic response by

CC chemokines, but not fMLP (48, 49). Alternatively, PLA2 inhibitors blocked both monocyte polarization and chemotaxis (48). Antisense oligonucleotide for cPLA2 induced a concentration-dependent inhibition of monocyte chemotaxis to all the CC chemokines tested (MCP-1, MCP-2, RANTES, and MIP-1α). On the contrary, chemotactic response to two 'classical' chemotactic factors, fMLP and C5a, was not affected by this treatment (50). This finding shows that cPLA2 is indeed an important effector enzyme for chemokine-induced monocyte migration. Both CC and CXC chemokines activate Pyk2/RAFTK tyrosine kinase and the Ras/Raf/MAP kinase pathways (51–53). It is important to note that MAP kinases (ERKs, p38, and JNK) are upstream of cPLA2 activation, and inhibitors of MAP kinases have been shown to reduce the chemotactic response to MCP-1 (52). There is also evidence for a role of phosphatidylinositol 3-kinase (PI3K) in chemokine receptor signal transduction (54). PI3K is activated by interaction with βγ subunit of eterotrimeric G proteins or by low molecular weight G proteins. Using CXCR1-transfected pre-B cells, IL-8 was shown to activate Rho, a low molecular weight G protein, and this effect was implicated in IL-8-induced adhesion to fibrinogen (55).

4. Chemokine functions

The main eponymous function of chemokines is to elicit directional migration of cells, mainly leukocytes, along a concentration gradient. The one notable exception to this rule is platelet factor 4 (PF4), the first molecularly cloned chemokine. PF4 is stored in the α granules of platelets; a canonical seven transmembrane domain PF4 receptor has not been identified and there is evidence that this molecule may exert anti-angiogenic activity. Chemokines are poor activators of leukocyte effector functions, including oxidative burst, granule release, and cytokine production. However, they activate cell functions related to the migration programme, including integrin avidity and expression of enzymes important for progression *in vivo*.

Chemokines affect the function of certain non-haematopoietic elements including endothelial cells and fibroblasts. In particular, CXC chemokines have been suggested to regulate angiogenesis. Molecules with an ELR motif preceding the CXC tandem (e.g. IL-8) have angiogenic activity whereas non-ELR CXC chemokines (e.g. IP-10, PF4) are anti-angiogenic. MCP-1 has recently been shown to promote collagen deposition by fibroblasts, an activity with considerable implications in diverse human disorders.

Some chemokines affect proliferation of haematopoietic precursors. In particular MIP-1α inhibits immature precursors whereas several chemokines have CSF-like stimulatory activity on more mature elements.

5. Regulation of chemokine production

Probably all cell types can produce chemokines under appropriate conditions. Two general modes of chemokine production can be defined (Table 2). Molecules such as SDF1 or MDC are produced constitutively either by specialized cells and organs, macrophages, dendritic cells, thymus, and lymphoid organs for MDC, or in a more

Table 2 Two modes of production of chemokines

Type of production	Family	Molecule	Main stimuli[a]	Main source
Tonic	CC	MDC	–	Thymus, dendritic cells, Mφ
		ELC	–	Lymphoid organs
		LARC/Exodus/MIP-3α	–	Thymus
	CXC	SDF1	–	Various
Inducible	CC	MCP-1 etc.	Microbial, inflammatory, immune	Diverse monocytes, endothelial cells, etc.
	CXC	IL-8 etc.	Microbial, inflammatory, immune	Diverse monocytes, endothelial cells, etc.
		IP-10 etc.	Immune (IFN-γ)	T cells, monocytes, etc.
	C	Lymphotactin	Immune	Class I restricted T cells
	CX3C	Fractalkine	Inflammatory	Endothelial cells

[a] Inflammatory: IL-1, TNF, IL-6, thrombin. Immune: IFN-γ, IL-4, IL-13, T cell receptor engagement.

diffuse way as for SDF1. However, most chemokines are produced upon cell activation. Interestingly, chemokines are also produced in a redundant way (polyspeirism, πολυσ, many, σπειρω, make). Usually the same cell produces many chemokines concomitantly in response to the same stimulus (Fig. 2 and Table 2). Polyspeirism is particularly striking for mononuclear phagocytes and endothelial cells exposed to bacterial products or primary inflammatory cytokines (Fig. 2). In these cells, bacterial lipopolysaccharide, IL-1, and TNF elicit production of MCPs, MIPs, RANTES (CC), fractalkine (CXC3C), and various CXC molecules (20, 24, 25, 36, 56).

The cytokine and cellular context dictates the pattern of chemokine production. For instance, certain stimuli (IL-4, IL-13, IL-10) have different or divergent effects in mononuclear phagocytes (Mφ) versus endothelial cells (EC). Furthermore, IFN-γ promotes the production of MCPs, active on monocytes and other mononuclear cells, and inhibits that of CXC chemokine with an ELR motif, active on neutrophils (Fig. 3).

6. Regulation of receptor expression during activation and deactivation of mononuclear phagocytes

Leukocyte infiltration into tissues is regulated by local production of chemotactic signals. Chemokine receptors are expressed on different types of leukocytes. Some receptors are restricted to certain cells (e.g. CXCR1 on PMN, and CCR3 on eosinophils and basophils), whereas others, such as CCR1 and CCR2, are expressed on different types of leukocytes. In addition, chemokine receptors are constitutively expressed on some cells, whereas they are inducible in others. Regulation of chemokine receptors is emerging as an alternative mechanism to control the level and the specificity of leukocyte migration. IL-2-activated, but not resting, T lymphocytes and NK cells migrate in response to MCPs. CXCR3 is expressed only in IL-2-activated T

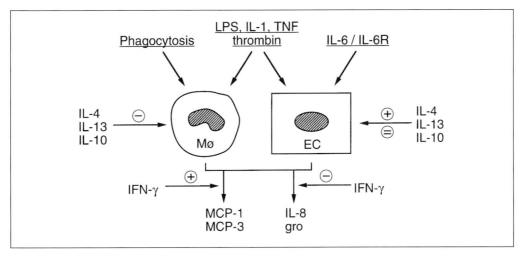

Fig. 3 Role of the cellular and cytokine context in skewing chemokine production. The cytokine and cellular context dictates the pattern of chemokine production in mononuclear phagocytes and endothelial cells. Certain stimuli (IL-4, IL-13, IL-10) have different or divergent effects in mononuclear phagocytes (Mϕ) versus endothelial cells (EC): + stimulation; – inhibition; = no effect. IFN-γ promotes the production of MCPs, active on monocytes and other mononuclear cells, and inhibits that of CXC chemokine with an ELR motif, active on neutrophils.

lymphocytes and IL-2 up-regulates CCR6 expression (21). In PMN, IL-8 receptors can be up-regulated or down-regulated by G-CSF and LPS, respectively (57).

It was recently described that inflammatory and anti-inflammatory agonists regulated in opposite ways CC chemokine receptor expression in human monocytes. LPS and other microbial agents caused a rapid and drastic reduction of CCR2 mRNA levels. The rate of nuclear transcription of CCR2 was not affected by LPS, whereas the mRNA half-life was reduced. The effect was drastic and rapid with an ED50 of ≈ 1 ng/ml and the half-maximal effect was reached with an optimal dose in ≈ 45 min. Inhibition of MCP-1 receptor expression was functionally relevant since LPS-treated monocytes showed a reduced capacity to bind and respond to MCP-1 chemotactically. The action of LPS on CC chemokine receptors was specific in that CXCR2 was unaffected (58). IFN-γ, a potent activator of mononuclear phagocytes, also inhibited CCR2 expression in a rapid (one hour) and selective manner by a reduction of the half-life of mRNA. Its effect was synergistic with the action of other pro-inflammatory molecules, such as IL-1, TNF, and LPS (59).

Conversely, incubation of human monocytes with IL-10 increased the expression of CCR1, 2, and 5 as evaluated by Northern blot analysis. No major variations in the expression of CXCR2 were detectable whereas CXCR4 mRNA levels were reduced (60). The effect of IL-10 was concentration dependent (EC50 = 0.3 ng/ml), and fast; already detectable after 30 minutes and reaching a plateau after two hours of stimulation. The estimated half-life of CCR5 mRNA was doubled after exposure to IL-10. In contrast, the rate of nuclear transcription of the gene, as investigated by nuclear run-off analysis, was not affected. Accordingly, IL-10-treated monocytes responded better to CC chemokines in terms of chemotactic migration and intracellular calcium transients and the effect was best observed when suboptimal agonist concentrations were used. IL-10-treated monocytes were also more easily infected by the macrophage tropic HIV strain BAL (60). This result is consistent with the use of CCR5 as major fusion co-receptor by BAL.

Regulation of receptor expression, in addition to agonist production, is probably a crucial point in the regulation of the chemokine system. An emerging paradigm indicates that at least some pro- and anti-inflammatory molecules exert reciprocal and opposing influences on chemokine agonist production and receptor expression. We speculate that the divergent effect of certain pro-inflammatory signals on agonist versus receptor expression may serve to retain mononuclear phagocytes at sites of inflammation, to prevent their reverse transmigration, and possibly, to limit excessive recruitment.

7. Regulation of receptor expression: dendritic cells

Dendritic cells (DC) have the unique capacity to initiate primary and secondary immune responses. They take up antigens in peripheral tissues and migrate to lymphoid organs where they present processed peptides to T cells. During migration DC undergo maturation from a 'processing' to a 'presenting' functional phenotype, characterized by the expression of co-stimulatory molecules, cytokine production, and high ability to stimulate T cell proliferation. For their central role in the regulation of immunity, DC are considered interesting tools and targets for immunotherapeutic intervention.

The response of DC to chemokines has been extensively characterized *in vitro*. Classical chemotactic agonists, such as formulated peptides and C5a, and some chemokines, induce directional migration and calcium fluxes of monocyte-derived DC (mono-DC) and CD34$^+$ cell-derived DC (CD34-DC) (61, 62). They include MCP-3, MCP-4, RANTES, MIP-1α, MIP-1β, and MIP-5 (CC) and SDF-1 (CXC). On the contrary, IL-8, Groβ, and IP-10, three members of the CXC chemokines, and eotaxin (CC) and lymphotactin (C) were inactive. PAF, a weak chemotactic factor for neutrophils and monocytes, efficiently induced chemotaxis and calcium transients in both mono-DC and CD34-DC (63). MDC (macrophage-derived chemokine), a new CC chemokine preferentially activated DC (two logs more potent on DC than on mononuclear cells) (64). Mono-DC express high levels of mRNA for CCR1, CCR2, and CCR5 receptors. Among the CXC chemokine receptors investigated, CXCR1, CXCR2, and CXCR4 were also expressed in DC (62). A similar pattern of chemokine receptor expression was observed in CD34-DC (Sozzani and Allavena, unpublished) with CCR6 being the only exception. This receptor is selectively expressed in CD34-DC that also migrate in response to MIP-3α/Larc/Exodus, the CCR6 ligand (65). Active chemokines increase the adhesion of mono-DC to endothelial cells and promote their transmigration across the endothelial cell monolayer. The latter response is inhibited by blocking antibodies to β1 integrins and by anti-CD31 (66). On the contrary, brief exposure of mono-DC to chemokines did not affect their ability to uptake FITC–

dextran or to promote allogenic mixed lymphocyte reaction (62). These results were generated using DC cultured from blood monocytes in the presence of granulocyte-macrophage colony-stimulating factor (GM-CSF) and IL-4/IL-13, or from CD34$^+$ cells in the presence of GM-CSF, stem cell factor, and TNF that can be considered to be at an immature stage. Maturation of DC can be induced *in vitro* by a variety of factors, including bacterial lipopolysaccharide, and the inflammatory cytokines TNF and IL-1, or the engagement of CD40 (67). Exposure of DC to inflammatory agonists, such as LPS, IL-1, and TNF, or culture in the presence of CD40 ligand, induced a rapid inhibition of chemotactic response to MIP-1α, MIP-1β, MIP-3α, RANTES, MCP-3, and fMLP (68, 69). As previously observed in phagocytes (57, 58), inhibition of chemotaxis was followed, with slower kinetics, by the reduction of membrane receptors and the down-regulation of mRNA receptor expression (68–71). Concomitantly, the expression of CCR7 and the migration to its ligand ELC/MIP-3β, a chemokine constitutively expressed in lymphoid organs, were strongly up-regulated, with a maximal effect at 24 hours. Up-regulation of CCR7 in DC migrating to secondary lymphoid organs seem to be of biological relevance, since *in situ* hybridization analysis has shown that ELC/MIP-3β is specifically expressed in areas rich in T cells in tonsils and spleen, where mature DC home, and becoming inter-digitating DC (69, 72). In contrast, MIP3α is produced at sites of inflammation (e.g. inflamed epithelial crypts of tonsils) where immature DC are recruited (69).

In vivo intradermal administration of GM-CSF leads to an increased number of DC within the human dermis. TNF, and possibly other LPS-induced cytokines, quickly recruit DC in the airway epithelia in a model of respiratory infection, and systemic administration of LPS induces a profound loss of MHC class II$^+$ cells from heart and kidney in the mouse (73). Since these pro-inflammatory agonists are inactive in promoting cell migration *in vitro*, it is likely that the observed DC mobilization observed *in vivo* is the result of secondary mediators, such as chemokines (62).

Finally, DC in the mucosal epithelium are likely to be the initial target for HIV-1 and, following migration to the nearest lymphoid station, contribute to the spreading of the virus. DC generated *in vitro*, as well as Langherans cells, express, and can be infected through, CCR5 and CXCR4, the two major fusion co-receptors for HIV macrophage and T cell tropic strains, respectively (74, 75).

8. Pathology

Chemokines are implicated in the pathophysiology of diverse disorders ranging from infectious diseases, immunoinflammatory disorders, and neoplasia. In particular, the identification of chemokine receptors as long sought fusion co-receptors for HIV has provided tremendous impetus to research in the field. A comprehensive review of the involvement of chemokines in disease is impossible given the space limitations of this review. For instance, elevated levels of MCP-1 have been detected in animal models of inflammatory and immune reactions including experimental allergic encephalomyelitis, pulmonary fibrosis, pulmonary granuloma, and renal ischaemia and reperfusion injury (for review see ref. 24).

In humans, MCP-1 elevations have been demonstrated in idiopathic pulmonary fibrosis, chronic active hepatitis, skin hypersensitivity reactions, rheumatoid arthritis, and atherosclerosis. In disorders such as lupus nephritis and in meningitis associated with HIV infection, there is evidence that measurement of this chemokine, taken as an example most familiar to the authors, may be of value in the diagnosis and follow-up of these disorders (76, 77).

In relation to pathology, here we will first review emerging information which indicates that chemokines and their receptors are an integral component of the Th1–Th2 paradigm of polarized immune responses. In addition, we will discuss involvement of chemokines in neoplastic disorders, which, we surmise, represent a paradigm for the *in vivo* function of these molecules.

8.1 Polarized Th1 and Th2 responses

T cells can be subdivided in polarized Type I and Type II cells, depending on the spectrum of cytokines that they are able to produce. T helper 1 cells (Th1 cells) are characterized by production by of TNF and IFN-γ, and activate immunity based on macrophage activation and effector functions. At the other extreme of the spectrum, Th2 cells are characterized by IL-4 and IL-5 production, and elicit immune responses based on the effector function of mastocytes and eosinophils. The latter cell types are typically involved in allergic inflammation.

Recent results indicate that chemokines are part of the Th1 and Th2 paradigm. It was found that polarized Th1 and Th2 populations differentially express chemokine receptors. In particular, as summarized in Table 3, Th1 cells characteristically express high levels of CCR5 and CXCR3 whereas Th2 cells express CCR4, CCR8 and, to a lesser degree, CCR3. In accordance with receptor expression, polarized Th1/Th2 cells (as well as CD8$^+$ T cells with a similar cytokine profile, unpublished observations) differentially respond to appropriate agonists for these receptors, including, for Th1 cells, MIP-1β and IP-10, and for Th2 cells, MDC, I309, and eotaxin (78, 79). Production of IP-10 and similar CXCR3 agonists such as ITAC, is induced by IFN-γ.

Table 3 Differential expression of chemokine receptors in human Th1 and Th2 cells

Receptor[a]	Prototypic agonists	Th1	Th2
CCR3	Eotaxin	–	±
CCR4	MDC	–	+++
CCR5	MIP-1β	+++	±
CCR8	1309	–	+
CXCR3	IP-10	+++	±

[a] Other receptors were similar (CCR1, CCR2, CXCR4) or not expressed (CXCR1 and CXCR2). Emerging evidence indicates that polarized Th1 and Th2 cells are different in chemokine receptor expression and responsiveness (4–6). They could be part of an amplification circuit of polarized responses.

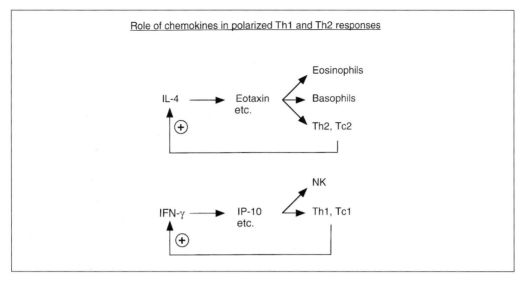

Role of chemokines in polarized Th1 and Th2 responses

Fig. 4 Chemokines in polarized Th1 and Th2 cells. Chemokines are part of amplification loops of polarized Th1 and Th2 responses.

Conversely, production of eotaxin and/or MDC is induced by IL-4 and IL-13, which are typical Th2 cytokines. Thus, as outlined in Fig. 4, chemokines are probably an essential part of an amplification circuit of polarized Th1 and Th2 responses.

8.2 Tumours as a paradigm for the *in vivo* function of chemokines

8.2.1 Chemokines and leukocyte infiltration

Analysis of mechanisms of recruitment of macrophages in tumours was one pathway that led to the identification of MCP-1 (5, 8, 12, 16, 80). Several lines of evidence suggest that MCP-1 can represent an important determinant of the levels of tumour-associated macrophages (TAM) (8, 80). In early studies with murine tumours or human tumours in nude mice a correlation was found between MCP-1 activity and percentage of TAM, a finding confirmed in subsequent experiments with the MCP-1 probe (81). Subcutaneous inoculation of tumour-derived human MCP-1, MCP-2, and MCP-3 led to macrophage infiltration (14, 82, 83). Finally, and conclusively, transfer of the mouse or human MCP-1 gene was associated with augmented levels of macrophage infiltration (84, 85). High expression of MCP-1 was associated with abrogation of tumourigenicity of CHO cells (84) but not of malignant mouse tumours (85). At low tumour inocula, MCP-1 gene transfer was associated with higher tumourigenicity and lung colonizing ability, despite a lower growth of resulting lesions (85, 86). These findings were interpreted in the light of the dual influence that TAM can exert on tumour growth (8, 80). Expression of MCP-1/JE was detected in various rat tumours (87). Using markers selective for monocyte-derived versus tissue macro-phages, MCP-1/JE gene transfer caused recruitment of mononuclear phagocytes from the blood compartment (87). Evidence for a role of MCP-1 in recruitment was obtained in human tumour xenografts (88).

Leukocyte infiltration in murine tumours is associated with administration of cytokines such as interferons, IL-2, or IL-4, by conventional routes or following gene transfer. Interferon, IL-12, and IL-2 induce endogenous chemokines in renal and colon cancer model (89, 90). Thus, secondary induction of chemokines may play a pivotal role in leukocyte recruitment in tumours treated with cytokines other than chemokines.

Various human tumour lines express MCP-1 *in vitro* spontaneously or after exposure to inflammatory signals, and some do so *in vivo*. The latter include gliomas, histiocytomas, sarcomas, and melanoma (12, 14, 91, 92). Expression of MCP-1 was recently found in Kaposi's sarcoma (KS) *in vivo* and in KS-derived spindle cell cultures (93). Since KS is characterized by a conspicuous macrophage infiltrate and is believed to represent a cytokine-propelled disease, production of MCP-1 may be particularly significant in this disease. Interestingly, human herpesvirus 8, which is probably involved in the pathogenesis of KS, encodes a constitutively active chemokine receptor that stimulates cell proliferation (94).

Human tumour lines of epithelial origin (breast, colon, ovary) (5, 7) release small molecular weight chemoattractant(s). Ovarian carcinoma is the one human tumour that has been most extensively studied for cytokine circuits between tumour cells and infiltrating leukocytes (95). These studies have also led to the design of thera-peutic strategies targeted to TAM, which gave encouraging results (96, 97). The role of chemokines in the ping-pong interaction between ovarian carcinoma cells and macrophages has been discussed (98). Freshly isolated ovarian carcinoma cells, primary cultures, and some established cell lines were shown in early studies to release tumour-derived chemotactic factor (TDCF) activity (5–7). These observations were recently revised (99). Immunohistochemistry and *in situ* hybridization have demonstrated that ovarian carcinoma cells and, in some tumours, stromal elements express MCP-1. High levels of MCP-1 were measured in the ascites (but not in blood) of patients with ovarian cancer but not in the peritoneal fluid of patients with non-malignant conditions. Production of MCP-1 and recruitment of TAM is likely to play an important role in progression of this disease because macrophage-derived cytokines promote the growth of ovarian carcinoma and its secondary implantation in peritoneal organs (100).

The expression of MCP-1 in relation to cervical cancer has been investigated by *in vitro* and *in vivo* approaches (83, 101). Somatic cell hybrids were generated between human papillomavirus type 18 (HPV18) cells and normal cells. Only non-tumouri-genic hybrids expressed MCP-1, whereas it was undetectable in tumourigenic segre-gants and in HPV positive cervical carcinoma lines. MCP-1, by *in situ* hybridization, was detected in certain human cervical cancers. In high grade squamous intra-epithelial lesions, MCP-1 expression was detected in normal, displastic, and neoplastic epithelia, as well as in macrophages and endothelial cells. MCP-1 expression was most prominent at epithelial–mesenchymal function and associated with macro-phage infiltration. In intraepithelial lesions, expression of MCP-1 and of the HPV oncogenes E6/E7 tended to be mutually exclusive, whereas, in squamous cell carcinoma, MCP-1 was expressed in the presence of transcriptional activity of E6/E7.

Various human tumours express chemokines of the CXC family. Some neoplasms of the melanocyte lineage express groα and the related molecule, IL-8, which induce proliferation and migration of melanoma cells (102–104). Transfer of the *gro* gene in an untransformed melanocytic line rendered it competent to form tumours in immunodeficient mice (102): this effect could be related to direct growth stimulation or to promotion of an inflammatory reaction, which would in turn favour tumour formation (see below). Inflammation and wound healing have in fact been im-plicated in initial steps of melanocyte oncogenesis (105, 106). IL-8 is produced by various human tumour lines *in vitro*, in particular carcinomas and brain tumours, either spontaneously or after exposure to IL-1 and TNF (107, 108). It has been speculated that IL-8 may contribute to lymphocytic infiltration in brain tumours (107). In addition, IL-8 has angiogenic activity (109), and could thus contribute to tumour angiogenesis. A novel member of the CXC family, GCP-2, was recently identified in supernatants of stimulated sarcoma cells (110). Direct evidence that ELR^+ CXC chemokines play a positive role in tumour angiogenesis has been obtained in non-small cell lung cancer (111, 112). In this human tumour, angiogenesis appears to be regulated by a balance between pro- and anti-angiogenic (IP-10) chemokines (111).

8.2.2 Defective systemic immunity and inflammation in cancer patients: a role for chemokines?

In terms of mounting immuno-inflammatory reactions, neoplastic disorders con-stitute, in a way, a paradox. As discussed above, many, if not all, tumours produce chemoattractants and are infiltrated by leukocytes: yet, it has long been known that neoplastic disorders are associated with immunosuppression and defective capacity to mount inflammatory reactions at sites other than the tumour (113). It was repeatedly demonstrated that circulating monocytes from cancer patients have a defective capacity to respond to chemoattractants (114–120). We have speculated that tumour-derived chemokines may play a role in the two seemingly contradictory aspects of monocyte function in neoplasia (Fig. 5) (100). Chemokines released con-tinuously from a growing tumour, may, beyond a certain tumour size, leak into the systemic circulation. Chemoattractants are classically known to cause desensitiz-ation, which, depending on the time of exposure, can be restricted to agents that use the same receptor (homologous) or involve other seven transmembrane domain receptors. Therefore, continuous exposure to tumour-derived chemoattractants may paralyse leukocytes. A more subtle anti-inflammatory action of chemoattractants depends on their effect on the receptors of the primary cytokines IL-1 and TNF (121, 122). Chemotactic agents cause rapid shedding of the TNF receptors, most efficiently RII/p75, and of Type II IL-1 'decoy' receptor.

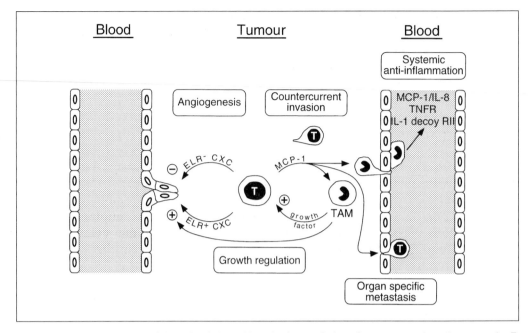

Fig. 5 A schematic overview of the role of chemokines in the regulation of tumour growth and metastasis. T, tumour cell; TAM, tumour-associated macrophages; ELR⁺ and ELR⁻ CXC, CXC chemokines with or without the ELR motif; TNFR and IL-1 RII, TNF and IL-1 receptor. MCP-1 is indicated as a prototypic CC chemokine. Countercurrent invasion as well as direct attraction for tumour cells may explain a positive role for MCP-1 in metastasis under certain conditions.

IL-8 was also able to induce rapid decoy RII release, though less efficiently than FMLP or C5a (121). However, IL-8 had additive effects with other elements in the cascade of recruitment such as platelet activating factor (Orlando, S., unpublished data). Most likely, rapid shedding of the TNFR and of the IL-1 decoy R serves to buffer primary pro-inflammatory cytokines leaking from sites of inflammation. Consistent with the concept of an anti-inflammatory potential of chemokines, systemic IL-8 inhibits local inflammatory reactions, and transgenic mice over-expressing MCP-1 have impaired resistance to intracellular pathogens (123). Thus, chemoattractant-induced continuous release of these molecules able to block IL-1 and TNF may contribute to a defective capacity to mount an inflammatory response systemically, coexisting with continuous leukocyte recruitment at the tumour site.

8.2.3 Monocyte functions other than chemotaxis

CC chemokines, and MCP-1 in particular, affect several functions of mononuclear phagocytes related to recruitment or to effector activity. Interaction and localized digestion of extracellular matrix components is essential for phagocyte extravasation and progression in tissues. MCP-1 induces production of gelatinase and of urokinase-type plasminogen activator (uPA) (124, 125). Concomitantly, MCP-1 augments expression of the cell surface receptor for uPA (uPA-R). Induction of gelatinase was also observed with MCP-2 and -3 (82). Thus, CC chemokines arm monocytes with the molecular tools that allow localized and polarized digestion of extracellular matrix components during recruitment. In tumour tissues, the release of lytic enzymes by MCP-1-stimulated TAM may provide a ready-made pathway for invasion of tumour cells (counter-current invasion) and thus contribute to augmented metastasis associated with inflammation (80, 82, 124–126). Accordingly, in one mouse model MCP-1 gene transfer augmented lung colony formation, (86) and chemokines are involved in selective metastasis to the kidney in a mouse lymphoma (127).

MCP-1 induces a respiratory burst in human monocytes, though it is a weak stimulus compared with other agonists (14, 128). Natural MCP-1 was reported to induce IL-1 and IL-6 but not TNF production (128). In another study, recombinant MCP-1 had little effect on IL-6 release (M. Sironi, *et al.* unpublished data). Human MCP-1 induced monocyte cytostasis for a tumour line (14) or synergized with bacterial products (but not with interferon-γ, IFN-γ) in the stimulation of mouse macrophage cytotoxicity (129, 130). In an interesting and intriguing study, human MCP-1 inhibited induction of the NO synthase in the macrophage cell line J774 (131). TAM have reduced NO synthase activity (132). If confirmed, this finding would suggest that MCP-1 could account for both recruitment and concomitant partial functional deactivation of TAM.

8.2.4 Overview

Chemokines play a dual role in the regulation of tumour growth and metastasis. Certain chemokines are produced by tumour cells and, by attracting macrophages and endothelial cells, provide optimal conditions for tumour growth and progression. We also speculate that chemokines leaking from sites of tumour growth may contribute to systemic impairment of the capacity to mount immune and inflammatory reactions frequently observed in advanced neoplasia. Conversely, the anti-angiogenic activity of chemokines and their capacity to recruit and activate immunocompetent cells can be exploited therapeutically in gene transfer studies.

A better understanding of the physiological role of chemokines in directing the traffic of dendritic cells and NK cells, crucial for the activation and orientation of specific immunity, may provide a basis for less empirical design of chemokine-based therapeutic strategies.

References

1. Butcher, E. C. (1991) Leukocyte-endothelial cell recognition: three (or more) steps to specificity and diversity. *Cell*, **67**, 1033.
2. Mantovani, A., Bussolino, F., and Dejana, E. (1992) Cytokine regulation of endothelial cell function. *FASEB J.*, **6**, 2591.
3. Ward, P. A., Remold, H. G., and David, J. R. (1970) The production by antigen-stimulated lymphocytes of a leukotactic factor distinct from migration inhibitory factor. *Cell. Immunol.*, **1**, 162.

4. Meltzer, M. S., Stevenson, M. M., and Leonard, E. J. (1977) Characterization of macrophage chemotaxis in tumor cell cultures and comparison with lymphocyte-derived chemotactic factors. *Cancer Res.*, **37**, 721.

5. Bottazzi, B., Polentarutti, N., Acero, R., Balsari, A., Boraschi, D., Ghezzi, P., *et al.* (1983) Regulation of the macrophage content of neoplasms by chemoattractants. *Science*, **220**, 210.

6. Bottazzi, B., Polentarutti, N., Balsari, A., Boraschi, D., Ghezzi, P., Salmona, M., *et al.* (1983) Chemotactic activity for mononuclear phagocytes of culture supernatants from murine and human tumor cells: evidence for a role in the regulation of the macrophage content of neoplastic tissues. *Int. J. Cancer*, **31**, 55.

7. Bottazzi, B., Ghezzi, P., Taraboletti, G., Salmona, M., Colombo, N., Bonazzi, C., *et al.* (1985) Tumor-derived chemotactic factor(s) from human ovarian carcinoma: evidence for a role in the regulation of macrophage content of neoplastic tissues. *Int. J. Cancer*, **36**, 167.

8. Mantovani, A., Bottazzi, B., Colotta, F., Sozzani, S., and Ruco, L. (1992) The origin and function of tumor-associated macrophages. *Immunol. Today*, **13**, 265.

9. Valente, A. J., Fowler, S. R., Sprague, E. A., Kelley, J. L., Suenram, C. A., and Schwartz, C. J. (1984) Initial characterization of a peripheral blood mononuclear cell chemoattractant derived from cultured arterial smooth muscle cells. *Am. J. Pathol.*, **117**, 409.

10. Zullo, J. N., Cochran, B. H., Huang, A. S., and Stiles, C. D. (1985) Platelet-derived growth factor and double-stranded ribonucleic acids stimulate expression of the same genes in 3T3 cells. *Cell*, **43**, 793.

11. Rollins, B. J., Morrison, E. D., and Stiles, C. D. (1988) Cloning and expression of JE, a gene inducible by platelet-derived growth factor and whose product has cytokine-like properties. *Proc. Natl. Acad. Sci. USA*, **85**, 3738.

12. Yoshimura, T., Robinson, E. A., Tanaka, S., Appella, E., Kuratsu, J., and Leonard, E. J. (1989) Purification and amino acid analysis of two human glioma-derived monocyte chemoattractants. *J. Exp. Med.*, **169**, 1449.

13. Matsushima, K., Larsen, C. G., DuBois, G. C., and Oppenheim, J. J. (1989) Purification and characterization of a novel monocyte chemotactic and activating factor produced by a human myelomonocytic cell line. *J. Exp. Med.*, **169**, 1485.

14. Zachariae, C. O., Anderson, A. O., Thompson, H. L., Appella, E., Mantovani, A., Oppenheim, J. J., *et al.* (1990) Properties of monocyte chemotactic and activating factor (MCAF) purified from a human fibrosarcoma cell line. *J. Exp. Med.*, **171**, 2177.

15. Van Damme, J., Decock, B., Lenaerts, J. P., Conings, R., Bertini, R., Mantovani, A., *et al.* (1989) Identification by sequence analysis of chemotactic factors for monocytes produced by normal and transformed cells stimulated with virus, double-stranded RNA or cytokine. *Eur. J. Immunol.*, **19**, 2367.

16. Graves, D. T., Jiang, Y. L., Williamson, M. J., and Valente, A. J. (1989) Identification of monocyte chemotactic activity produced by malignant cells. *Science*, **245**, 1490.

17. Furutani, Y., Nomura, H., Notake, M., Oyamada, Y., Fukuy, T., Yamada, M., *et al.* (1989) Cloning and sequencing of the cDNA for human monocyte chemotactic and activating factor (MCAF). *Biochem. Biophys. Res. Commun.*, **159**, 248.

18. Yoshimura, T., Yuhki, N., Moore, S. K., Appella, E., Lerman, M. I., and Leonard, E. J. (1989) Human monocyte chemoattractant proein-1 (MCP-1). Full-length cDNA cloning, expression in mitogen-stimulation blood mononuclear leukocytes, and sequence similarity to mouse competence gene JE. *FEBS Lett.*, **244**, 487.

19. Bottazzi, B., Colotta, F., Sica, A., Nobili, N., and Mantovani, A. (1990) A chemoattractant expressed in human sarcoma cells (tumor-derived chemotactic factor, TDCF) is identical

to monocyte chemoattractant protein-1/monocyte chemotactic and activating factor (MCP-1/MCAF). *Int. J. Cancer*, **45**, 795.

20. Schall, T. J. (1994) The chemokines. In *The cytokine handbook* (ed. A. Thomson), pp. 419–60. Academic Press, London.

21. Baggiolini, M., Dewald, B., and Moser, B. (1997) Human chemokines: an update. *Annu. Rev. Immunol.*, **15**, 675.

22. Hedrick, J. A. and Zlotnik, A. (1996) Chemokines and lymphocyte biology. *Curr. Opin. Immunol.*, **8**, 343.

23. Rollins, B. J. (1997) Chemokines. *Blood*, **90**, 909.

24. Mantovani, A., Allavena, P., Vecchi, A., and Sozzani, S. (1998) Chemokines and chemokine receptors during activation and deactivation of monocytes and dendritic cells and in amplification of Th1 versus Th2 responses. *Int. J. Clin. Lab. Res.*, **28**, 77.

25. Ben-Baruch, A., Michiel, D. F., and Oppenheim, J. J. (1995) Signals and receptors involved in recruitment of inflammatory cells. *J. Biol. Chem.*, **270**, 11703.

26. Kelner, G. S., Kennedy, J., Bacon, K. B., *et al.* (1994) Lymphotactin: a cytokine that represents a new class of chemokine. *Science*, **266**, 1395.

27. Bazan, J. F., Bacon, K. B., Hardiman, G., *et al.* (1997) A new class of membrane-bound chemokine with a CX3C motif. *Nature*, **385**, 640.

28. Pan, Y., Lloyd, C., Zhou, H., *et al.* (1997) Neurotactin, a membrane-anchored chemokine upregulated in brain inflammation. *Nature*, **387**, 611.

29. Bianchi, G., Sozzani, S., Zlotnik, A., Mantovani, A., and Allavena, P. (1996) Migratory response of human NK cells to lymphotactin. *Eur. J. Immunol.*, **26**, 3238.

30. Imai, T., Hieshima, K., Haskell, C., *et al.* (1997) Identification and molecular character- ization of fractalkine receptor CX3CR1, which mediates both leukocyte migration and adhesion. *Cell*, **91**, 521.

31. Izumi, S., Hirai, K., Miyamasu, M., *et al.* (1997) Expression and regulation of monocyte chemoattractant protein-1 by human eosinophils. *Eur. J. Immunol.*, **27**, 816.

32. Legler, D. F., Loetscher, M., Roos, R. S., Clarklewis, I., Baggiolini, M., and Moser, B. (1998) B cell-attracting chemokine 1, a human CXC chemokine expressed in lymphoid tissues, selectively attracts B lymphocytes via BLR1/CXCR5. *J. Exp. Med.*, **187**, 655.

33. Nibbs, R. J. B., Wylie, S. M., Yang, J. Y., Landau, N. R., and Graham, G. J. (1997) Cloning and characterization of a novel promiscuous human beta-chemokine receptor D6. *J. Biol. Chem.*, **272**, 32078.

34. Yoshida, T., Imai, T., Kakizaki, M., Nishimura, M., Takagi, S., and Yoshie, O. (1998) Identification of single C motif-1 lymphotactin receptor XCR1. *J. Biol. Chem.*, **273**, 16551.

35. Oppenheim, J. J., Zachariae, C. O., Mukaida, N., and Matsushima, K. (1991) Properties of the novel proinflammatory supergene 'intercrine' cytokine family. *Annu. Rev. Immunol.*, **9**, 617.

36. Baggiolini, M., Dewald, B., and Moser, B. (1994) Interleukin-8 and related chemotactic cytokines—CXC and CC chemokines. *Adv. Immunol.*, **55**, 99.

37. Miller, M. D. and Krangel, M. S. (1992) Biology and biochemistry of the chemokines: a family of chemotactic and inflammatory cytokines. *Crit. Rev. Immunol.*, **12**, 17.

38. Sozzani, S., Locati, M., Zhou, D., *et al.* (1995) Receptors, signal transduction and spectrum of action of monocyte chemotactic protein-1 and related chemokines. *J. Leuk. Biol.*, **57**, 788.

39. Sozzani, S., Luini, W., Molino, M., Jílek, P., Bottazzi, B., Cerletti, C., *et al.* (1991) The signal transduction pathway involved in the migration induced by a monocyte chemotactic cytokine. *J. Immunol.*, **147**, 2215.

40. McColl, S. R., Hachicha, M., Levasseur, S., Neote, K., and Schall, T. J. (1993) Uncoupling of

early signal transduction events from effector function in human peripheral blood neutrophils in response to recombinant macrophage inflammatory proteins-1 alpha and -1 beta. *J. Immunol.*, **150**, 4550.

41. Heinrich, J. N., Ryseck, R. P., Macdonald-Bravo, H., and Bravo, R. (1993) The product of a novel growth factor-activated gene, fic, is a biologically active CC-type cytokine. *Mol. Cell. Biol.*, **13**, 2020.

42. Bischoff, S. C., Krieger, M., Brunner, T., and Dahinden, C. A. (1992) Monocyte chemotactic protein 1 is a potent activator of human basophils. *J. Exp. Med.*, **175**, 1271.

43. Myers, S. J., Wong, L. M., and Charo, I. F. (1995) Signal transduction and ligand specificity of the human monocyte chemoattractant protein-1 receptor in transfected embryonic kidney cells. *J. Biol. Chem.*, **270**, 5786.

44. Sozzani, S., Molino, M., Locati, M., Luini, W., Cerletti, C., Vecchi, A., *et al.* (1993) Receptor-activated calcium influx in human monocytes exposed to monocyte chemotactic protein-1 and related cytokines. *J. Immunol.*, **150**, 1544.

45. Sozzani, S., Zhou, D., Locati, M., *et al.* (1994) Receptors and transduction pathways for monocyte chemotactic protein-2 and monocyte chemotactic protein-3—similarities and differences with MCP-1. *J. Immunol.*, **152**, 3615.

46. Wu, D., LaRosa, G. J., and Simon, M. I. (1993) G protein-coupled signal transduction pathways for interleukin-8. *Science*, **261**, 101.

47. Kuang, Y. N., Wu, Y. P., Jiang, H. P., and Wu, D. Q. (1996) Selective G protein coupling by CC chemokine receptors. *J. Biol. Chem.*, **271**, 3975.

48. Locati, M., Zhou, D., Luini, W., Evangelista, V., Mantovani, A., and Sozzani, S. (1994) Rapid induction of arachidonic acid release by monocyte chemotactic protein-1 and related chemokines—role of Ca^{2+} influx, synergism with platelet-activating factor and significance for chemotaxis. *J. Biol. Chem.*, **269**, 4746.

49. Sozzani, S., Rieppi, M., Locati, M., Zhou, D., Bussolino, F., Van Damme, J., *et al.* (1994) Synergism between platelet activating factor and CC chemokines for arachidonate release in human monocyte. *Biochem. Biophys. Res. Commun.*, **199**, 761.

50. Locati, M., Lamorte, G., Luini, W., Introna, M., Bernasconi, S., Mantovani, A., *et al.* (1996) Inhibition of monocyte chemotaxis to CC chemokines by antisense oligonucleotide for cytosolic phospholipase A2. *J. Biol. Chem.*, **271**, 6010.

51. Knall, C., Young, S., Nick, J. A., Buhl, A. M., Worthen, G. S., and Johnson, G. L. (1996) Interleukin-8 regulation of the Ras/Raf/mitogen-activated protein kinase pathway in human neutrophils. *J. Biol. Chem.*, **271**, 2832.

52. Yen, H. H., Zhang, Y. J., Penfold, S., and Rollins, B. J. (1997) MCP-1-mediated chemotaxis requires activation of non-overlapping signal transduction pathways. *J. Leuk. Biol.*, **61**, 529.

53. Davis, C. B., Dikic, I., Unutmaz, D., Hill, C. M., Arthos, J., Siani, M. A., *et al.* (1997) Signal transduction due to HIV-1 envelope interactions with chemokine receptors CXCR4 or CCR5. *J. Exp. Med.*, **186**, 1793.

54. Turner, L., Ward, S. G., and Westwick, J. (1995) RANTES-activated human T lymphocytes —a role for phosphoinositide 3-kinase. *J. Immunol.*, **155**, 2437.

55. Laudanna, C., Campbell, J. J., and Butcher, E. C. (1996) Role of rho in chemoattractant-activated leukocyte adhesion through integrins. *Science*, **271**, 981.

56. Mantovani, A., Bussolino, F., and Introna, M. (1997) Cytokine regulation of endothelial cell function: from molecular level to the bed side. *Immunol. Today*, **18**, 231.

57. Lloyd, A. R., Biragyn, A., Johnston, J. A., *et al.* (1995) Granulocyte-colony stimulating factor and lipopolysaccharide regulate the expression of interleukin 8 receptors on polymorphonuclear leukocytes. *J. Biol. Chem.*, **270**, 28188.

58. Sica, A., Saccani, A., Borsatti, A., *et al.* (1997) Bacterial lipopolysaccharide rapidly inhibits expression of CC chemokine receptors in human monocytes . *J. Exp. Med.*, **185**, 969.

59. Penton-Rol, G., Polentarutti, N., Luini, W., Borsatti, A., Mancinelli, R., Sica, A., *et al.* (1998) Selective inhibition of expression of the chemokine receptor CCR2 in human monocytes by IFN-γ. *J. Immunol.*, **160**, 3869.

60. Sozzani, S., Ghezzi, S., Iannolo, G., *et al.* (1998) Interleukin-10 increases CCR5 expression and HIV infection in human monocytes. *J. Exp. Med.*, **187**, 439.

61. Sozzani, S., Sallusto, F., Luini, W., *et al.* (1995) Migration of dendritic cells in response to formyl peptides, C5a and a distinct set of chemokines. *J. Immunol.*, **155**, 3292.

62. Sozzani, S., Luini, W., Borsatti, A., *et al.* (1997) Receptor expression and responsiveness of human dendritic cells to a defined set of CC and CXC chemokines. *J. Immunol.*, **159**, 1993.

63. Sozzani, S., Longoni, D., Bonecchi, R., *et al.* (1997) Human monocyte-derived and CD34+ cell-derived dendritic cells express functional receptors for platelet activating factor. *FEBS Lett.*, **418**, 98.

64. Godiska, R., Chantry, D., Raport, C. J., Sozzani, S., Allavena, P., Leviten, D., *et al.* (1997) Human macrophage derived chemokine (MDC) a novel chemoattractant for monocytes, monocyte derived dendritic cells, and natural killer cells. *J. Exp. Med.*, **185**, 1595.

65. Power, C. A., Church, D. J., Meyer, A., *et al.* (1997) Cloning and characterization of a specific receptor for the novel CC chemokine MIP-3 alpha from lung dendritic cells. *J. Exp. Med.*, **186**, 825.

66. D'Amico, G., Bianchi, G., Bernasconi, S., Bersani, L., Piemonti, L., Sozzani, S., *et al.* (1998) Adhesion, transendothelial migration, and reverse transmigration of *in vitro* cultured dendritic cells. *Blood*, **92**, 207.

67. Bancherau, J. and Steinman, R. M. (1998) Dendritic cells and the control of immunity. *Nature*, **392**, 245.

68. Sozzani, S., Allavena, P., DAmico, G., *et al.* (1998) Cutting edge: differential regulation of chemokine receptors during dendritic cell maturation: a model for their trafficking properties. *J. Immunol.*, **161**, 1083.

69. Dieu, M. C., Vanbervliet, B., Vicari, A., *et al.* (1998) Selective recruitment of immature and mature dendritic cells by distinct chemokines expressed in different anatomic sites. *J. Exp. Med.*, **188**, 373.

70. Sallusto, F., Schaerli, P., Loetscher, P., Schaniel, C., Lenig, D., Mackay, C. R., *et al.* (1998) Rapid and coordinated switch in chemokine receptor expression during dendritic cell maturation. *Eur. J. Immunol.*, **28**, 2760.

71. Granelli-Piperno, A., Delgado, E., Finkel, V., Paxton, W., and Steinman, R. M. (1998) Immature dendritic cells selectively replicate macrophagetropic (M-tropic) human immunodeficiency virus type 1, while mature cells efficiently transmit both M- and T-tropic virus to T cells. *J. Virol.*, **72**, 2733.

72. Ngo, V. N., Tang, H. L., and Cyster, J. G. (1998) Epstein–Barr virus-induced molecule 1 ligand chemokine is expressed by dendritic cells in lymphoid tissues and strongly attracts naive T cells and activated B cells. *J. Exp. Med.*, **188**, 181.

73. Austyn, J. M. (1996) New insights into the mobilization and phagocytic activity of dendritic cells. *J. Exp. Med.*, **183**, 1287.

74. Granelli-Piperno, A., Moser, B., Pope, M., *et al.* (1996) Efficient interaction of HIV-1 with purified dendritic cells via multiple chemokine coreceptors. *J. Exp. Med.*, **184**, 2433.

75. Zaitseva, M., Blauvelt, A., Lee, S., Lapham, C. K., Klaus-Kovtun, V., Motowski, H., *et al.* (1997) Expression and function of CCR5 and CXCR4 on human Langerhans cells and macrophages: implication for HIV primary infection. *Nature Med.*, **3**, 1369.

76. Bernasconi, S., Cinque, P., Peri, G., *et al.* (1996) Selective elevation of monocyte chemotactic protein-1 in the cerebrospinal fluid of AIDS patients with cytomegalovirus encephalitis. *J. Infect. Dis.*, **174**, 1098.

77. Noris, M., Bernasconi, S., Casiraghi, F., Sozzani, S., Gotti, E., Remuzzi, G., *et al.* (1995) Monocyte chemoattractant protein-1 is excreted in excessive amounts in the urine of patients with lupus nephritis. *Lab. Invest.*, **73**, 804.

78. Bonecchi, R., Bianchi, G., Bordignon, P. P., *et al.* (1998) Differential expression of chemokine receptors and chemotactic responsiveness of type 1 T helper cells (Th1s) and Th2s. *J. Exp. Med.*, **187**, 129.

79. Sallusto, F., Mackay, C. R., and Lanzavecchia, A. (1997) Selective expression of the eotaxin receptor CCR3 by human T helper 2 cells. *Science*, **277**, 2005.

80. Opdenakker, G. and Van Damme, J. (1992) Chemotactic factors, passive invasion and metastasis of cancer cells. *Immunol. Today*, **13**, 463.

81. Walter, S., Bottazzi, B., Govoni, D., Colotta, F., and Mantovani, A. (1991) Macrophage infiltration and growth of sarcoma clones expressing different amounts of monocyte chemotactic protein/JE. *Int. J. Cancer*, **49**, 431.

82. Van Damme, J., Proost, P., Lenaerts, J. P., and Opdenakker, G. (1992) Structural and functional identification of two human, tumor-derived monocyte chemotactic proteins (MCP-2 and MCP-3) belonging to the chemokine family. *J. Exp. Med.*, **176**, 59.

83. Hirose, K., Hakozaki, M., Nyunoya, Y., *et al.* (1995) Chemokine gene transfection into tumour cells reduced tumorigenicity in nude mice in association with neutrophilic infiltration. *Br. J. Cancer*, **72**, 708.

84. Rollins, B. J. and Sunday, M. E. (1991) Suppression of tumor formation *in vivo* by expression of the JE gene in malignant cells. *Mol. Cell. Biol.*, **11**, 3125.

85. Bottazzi, B., Walter, S., Govoni, D., Colotta, F., and Mantovani, A. (1992) Monocyte chemotactic cytokine gene transfer modulates macrophage infiltration, growth, and susceptibility to IL-2 therapy of a murine melanoma. *J. Immunol.*, **148**, 1280.

86. Mantovani, A., Bottazzi, B., Sozzani, S., Peri, G., Allavena, P., Dong, Q. G., *et al.* (1993) Cytokine regulation of tumour-associated macrophages. *Res. Immunol.*, **144**, 280.

87. Yamashiro, S., Takeya, M., Nishi, T., Kuratsu, J., Yoshimura, T., Ushio, Y., *et al.* (1994) Tumor-derived monocyte chemoattractant protein-1 induces intratumoral infiltration of monocyte-derived macrophage subpopulation in transplanted rat tumors. *Am. J. Pathol.*, **145**, 856.

88. Melani, C., Pupa, S. M., Stoppacciaro, A., Menard, S., Colnaghi, M. I., Parmiani, G., *et al.* (1995) An *in vivo* model to compare human leukocyte infiltration in carcinoma xenografts producing different chemokines. *Int. J. Cancer*, **62**, 572.

89. Sonouchi, K., Hamilton, T. A., Tannenbaum, C. S., Tubbs, R. R., Bukowski, R., and Finke, J. H. (1994) Chemokine gene expression in the murine renal cell carcinoma, renca, following treatment *in vivo* with interferon-alpha and interleukin-2. *Am. J. Pathol.*, **144**, 747.

90. Tannenbaum, C. S., Wicker, N., Armstrong, D., Tubbs, R., Finke, J., Bukowski, R. M., *et al.* (1996) Cytokine and chemokine expression in tumors of mice receiving systemic therapy with IL-12. *J. Immunol.*, **156**, 693.

91. Graves, D. T., Barnhill, R., Galanopoulos, T., and Antoniades, H. N. (1992) Expression of monocyte chemotactic protein-1 in human melanoma *in vivo*. *Am. J. Pathol.*, **140**, 9.

92. Takeya, M., Yoshimura, T., Leonard, E. J., Kato, T., Okabe, H., and Takahashi, K. (1991) Production of monocyte chemoattractant protein-1 by malignant fibrous histiocytoma: relation to the origin of histiocyte-like cells. *Exp. Mol. Pathol.*, **54**, 61.

93. Sciacca, F. L., Stürzl, M., Bussolino, F., *et al.* (1994) Expression of adhesion molecules,

platelet-activating factor, and chemokines by Kaposi's sarcoma cells. *J. Immunol.*, **153**, 4816.

94. Arvanitakis, L., GerasRaaka, E., Varma, A., Gershengorn, M. C., and Cesarman, E. (1997) Human herpesvirus KSHV encodes a constitutively active G-protein-coupled receptor linked to cell proliferation. *Nature*, **385**, 347.

95. Burke, F., Relf, M., Negus, R., and Balkwill, F. (1996) A cytokine profile of normal and malignant ovary. *Cytokine*, **8**, 578.

96. Allavena, P., Peccatori, F., Maggioni, D., *et al.* (1990) Intraperitoneal recombinant gamma-interferon in patients with recurrent ascitic ovarian carcinoma: modulation of cytotoxicity and cytokine production in tumor-associated effectors and of major histocompatibility antigen expression on tumor cells. *Cancer Res.*, **50**, 7318.

97. Colombo, N., Peccatori, F., Paganin, C., Bini, S., Brandely, M., Mangioni, C., *et al.* (1992) Anti-tumor and immunomodulatory activity of intraperitoneal IFN-gamma in ovarian carcinoma patients with minimal residual tumor after chemotherapy. *Int. J. Cancer*, **51**, 42.

98. Evans, R. (1972) Macrophages in syngeneic animal tumours. *Transplantation*, **14**, 468.

99. Negus, R. P., Stamp, G. W., Relf, M. G., *et al.* (1995) The detection and localization of monocyte chemoattractant protein-1 (MCP-1) in human ovarian cancer. *J. Clin. Invest.*, **95**, 2391.

100. Mantovani, A. (1994) Tumor-associated macrophages in neoplastic progression: a paradigm for the *in vivo* function of chemokines. *Lab. Invest.*, **71**, 5.

101. Riethdorf, L., Riethdorf, S., Gutzlaff, K., Prall, F., and Loning, T. (1996) Differential expression of the monocyte chemoattractant protein-1 gene in human papillomavirus-16-infected squamous intraepithelial lesions and squamous cell carcinomas of the cervix uteri. *Am. J. Pathol.*, **149**, 1469.

102. Balentien, E., Mufson, B. E., Shattuck, R. L., Derynck, R., and Richmond, A. (1991) Effects of MGSA/GRO alpha on melanocyte transformation. *Oncogene*, **6**, 1115.

103. Wang, J. M., Taraboletti, G., Matsushima, K., Van Damme, J., and Mantovani, A. (1990) Induction of haptotactic migration of melanoma cells by neutrophil activating protein/interleukin-8. *Biochem. Biophys. Res. Commun.*, **169**, 165.

104. Schadendorf, D., Moller, A., Algermissen, B., Worm, M., Sticherling, M., and Czarnetzki, B. M. (1993) IL-8 produced by human malignant melanoma cells *in vitro* is an essential autocrine growth factor. *J. Immunol.*, **151**, 2667.

105. Mintz, B. and Silvers, W. K. (1993) Transgenic mouse model of malignant skin melanoma. *Proc. Natl. Acad. Sci. USA*, **90**, 8817.

106. Medrano, E. E., Farooqui, J. Z., Boissy, R. E., Boissy, Y. L., Akadiri, B., and Nordlund, J. J. (1993) Chronic growth stimulation of human adult melanocytes by inflammatory mediators *in vitro*: implications for nevus formation and initial steps in melanocyte oncogenesis. *Proc. Natl. Acad. Sci. USA*, **90**, 1790.

107. Van Meir, E., Ceska, M., Effenberger, F., *et al.* (1992) Interleukin-8 is produced in neoplastic and infectious diseases of the human central nervous system. *Cancer Res.*, **52**, 4297.

108. Sakamoto, K., Masuda, T., Mita, S., *et al.* (1992) Interleukin-8 is constitutively and commonly produced by various human carcinoma cell lines. *Int. J. Clin. Lab. Res.*, **22**, 216.

109. Koch, A. E., Polverini, P. J., Kunkel, S. L., Harlow, L. A., DiPietro, L. A., Elner, V. M., *et al.* (1992) Interleukin-8 as a macrophage-derived mediator of angiogenesis. *Science*, **258**, 1798.

110. Proost, P., De Wolf Peeters, C., Conings, R., Opdenakker, G., Billiau, A., and Van

Damme, J. (1993) Identification of a novel granulocyte chemotactic protein (GCP-2) from human tumor cells. *In vitro* and *in vivo* comparison with natural forms of GRO, IP-10, and IL-8. *J. Immunol.*, **150**, 1000.

111. Arenberg, D. A., Kunkel, S. L., Polverini, P. J., *et al.* (1996) Interferon-gamma-inducible protein 10 (IP-10) is an angiostatic factor that inhibits human non-small cell lung cancer (NSCLC) tumorigenesis and spontaneous metastases. *J. Exp. Med.*, **184**, 981.

112. Smith, D. R., Polverini, P. J., Kunkel, S. L., Orringer, M. B., Whyte, R. I., Burdick, M. D., *et al.* (1994) Inhibition of IL-8 attenuates angiogenesis in bronchogenic carcinoma. *J. Exp. Med.*, **179**, 1409.

113. Snyderman, R. and Cianciolo, G. J. (1984) Immunosuppressive activity of the retroviral envelope protein P15E and its possible relationship to neoplasia. *Immunol. Today*, **5**, 240.

114. Boechter, D. and Leonard, E. J. (1974) Abnormal monocyte chemotactic response in mice. *J. Natl. Cancer Inst.*, **52**, 1091.

115. Normann, S. J., Schardt, M., and Sorkin, E. (1981) Biphasic depression of macrophage function after tumor transplantation. *Int. J. Cancer*, **28**, 185.

116. Normann, S. J. and Sorkin, E. (1976) Cell-specific defect in monocyte function during tumor growth. *J. Natl. Cancer Inst.*, **57**, 135.

117. Normann, S. J. and Sorkin, E. (1977) Inhibition of macrophage chemotaxis by neoplastic and other rapidly proliferating cells *in vitro*. *Cancer Res.*, **37**, 705.

118. Stevenson, M. M. and Meltzer, M. S. (1976) Depressed chemotactic responses *in vitro* of peritoneal macrophages from tumor-bearing mice. *J. Natl. Cancer Inst.*, **57**, 847.

119. Snyderman, R. and Pike, M. C. (1976) An inhibitor of macrophage chemotaxis produced by neoplasms. *Science*, **192**, 370.

120. Cianciolo, G. J., Hunter, J., Silva, J., Haskill, J. S., and Snyderman, R. (1981) Inhibitors of monocyte responses to chemotaxins are present in human cancerous effusions and react with monoclonal antibodies to the P15(E) structural protein of retroviruses. *J. Clin. Invest.*, **68**, 831.

121. Colotta, F., Orlando, S., Fadlon, E. J., Sozzani, S., Matteucci, C., and Mantovani, A. (1995) Chemoattractants induce rapid release of the interleukin 1 type II decoy receptor in human polymorphonuclear cells. *J. Exp. Med.*, **181**, 2181.

122. Porteu, F. and Nathan, C. (1990) Shedding of tumor necrosis factor receptor by activated human neutrophils. *J. Exp. Med.*, **172**, 599.

123. Rutledge, B. J., Rayburn, H., Rosenberg, R., North, R. J., Gladue, R. P., Corless, C. L., *et al.* (1995) High level monocyte chemoattractant protein-1 expression in transgenic mice increases their susceptibility to intracellular pathogens. *J. Immunol.*, **155**, 4838.

124. Opdenakker, G. and Van Damme, J. (1992) Cytokines and proteases in invasive processes: molecular similarities between inflammation and cancer. *Cytokine*, **4**, 251.

125. Mantovani, A., Sozzani, S., Bottazzi, B., Peri, G., Sciacca, F. L., Locati, M., *et al.* (1993) Monocyte chemotactic protein-1 (MCP-1): signal transduction and involvement in the regulation of macrophage traffic in normal and neoplastic tissues. *Adv. Exp. Med. Biol.*, **351**, 47.

126. Opdenakker, G., Froyen, G., Fiten, P., Proost, P., and Van Damme, J. (1993) Human monocyte chemotactic protein-3 (MCP-3): molecular cloning of the cDNA and comparison with other chemokines. *Biochem. Biophys. Res. Commun.*, **191**, 535.

127. Wang, J. M., Chertov, O., Proost, P., *et al.* (1998) Purification and identification of chemokines potentially involved in kidney-specific metastasis by a murine lymphoma variant: induction of migration and NFkB activation. *Int. J. Cancer*, **75**, 900.

128. Jiang, Y., Beller, D. I., Frendl, G., and Graves, D. T. (1992) Monocyte chemoattractant

protein-1 regulates adhesion molecule expression and cytokine production in human monocytes. *J. Immunol.*, **148**, 2423.

129. Singh, R. K., Berry, K., Matsushima, K., Yasumoto, K., and Fidler, I. J. (1993) Synergism between human monocyte chemotactic and activating factor and bacterial products for activation of tumoricidal properties in murine macrophages. *J. Immunol.*, **151**, 2786.

130. Asano, T., An, T., Jia, S. F., and Kleinerman, E. S. (1996) Altered monocyte chemotactic and activating factor gene expression in human glioblastoma cell lines increased their susceptibility to cytotoxicity. *J. Leuk. Biol.*, **59**, 916.

131. Rojas, A., Delgado, R., Glaria, L., and Palacios, M. (1993) Monocyte chemotactic protein-1 inhibits the induction of nitric oxide synthase in J774 cells. *Biochem. Biophys. Res. Commun.*, **196**, 274.

132. DiNapoli, M. R., Calderon, C. L., and Lopez, D. M. (1996) The altered tumoricidal capacity of macrophages isolated from tumor-bearing mice is related to reduced expression of the inducible nitric oxide synthase gene. *J. Exp. Med.*, **183**, 1323.

6 | Cytokines and cytokine receptors encoded by viruses

JULIAN A. SYMONS AND GEOFFREY L. SMITH

1. Role of cytokines in antiviral defence

Cytokines are critical for the immunological control of viral infection. A number of cytokines, such as Type I interferon (IFN-α/β), Type II IFN (IFN-γ), and tumour necrosis factor-α (TNF-α), have been shown to directly suppress virus replication in the host. Others may act indirectly either by inducing antiviral cytokines, for example, interleukin (IL) 12 and IL-18 induce production of IFN-γ, or by activating the cellular and humoral immune responses that mediate virus clearance. The pivotal role of cytokines in antiviral defence has been demonstrated elegantly with the use of cytokine and cytokine receptor knock-out mice and by the attenuation of viral infection in virus recombinants expressing various cytokines. In addition, in recent years it has become apparent that many viruses have hijacked components of the cytokine system in order to evade or suppress the immune response mounted against them. This review will focus on the many cytokine-related immunomodulatory genes that are encoded by large DNA viruses.

2. Poxviruses and herpesviruses

Poxviruses are large complex viruses with linear double-stranded genomes ranging from 120–280 kb in size. Unusually for DNA viruses, they replicate in the cytoplasm of the cell and for this reason they encode a large number of enzymes involved in nucleotide biosynthesis, DNA replication, and transcription (1). These genes are essential for virus growth *in vitro* and are consequently highly conserved in poxviruses and are located in the central portion of the genome. Poxviruses are divided into two subfamilies, *Chordopoxvirinea* and *Entomopoxvirinea*, based on vertebrate and insect host range. The *Chorodpoxvirinea* are further subdivided into eight genera. Members of a genus are genetically and antigenically related and have a similar morphology.

The *orthopoxvirus*, variola and the *molluscipoxvirus*, molluscum contagiosum are the only poxviruses that infect only humans. Until its eradication in 1977, variola virus, the causative agent of smallpox, was the most important poxvirus to infect humans. Variola virus induced an acute, severe, generalized disease characterized by an extensive rash and fever. Mortality rates were as high as 40%. In the 1950s, 150 years after the introduction of vaccination, 50 million cases of smallpox still occurred each year. The outcome of infection was either death, or recovery and subsequent elimination of the virus. Recovery was associated with long-term immunity to reinfection. Jenner introduced vaccination against smallpox in 1796 using cowpox virus and, in 1801, predicted smallpox would be eradicated. Extensive vaccination with vaccinia virus (VV), a distinct orthopoxvirus, in the 20th century ensured Jenner's prophecy was fulfilled.

Because of their large coding capacity, poxviruses have many genes near the genomic termini that are non-essential for virus replication in tissue culture but aid virus replication within the host. These genes encode factors that affect virus host range, prevent apoptosis of the infected cell, block complement-mediated killing, and synthesize potentially immunosuppressive steroids. In addition, poxviruses express many proteins that target cytokines and cytokine receptors and use these molecules both as growth factors and for immune evasion (2).

The herpesvirus family is composed of a diverse, phylogenetically ancient family of large double-stranded DNA viruses with genomes between 120–230 kb (3). Herpesviruses are classified into three main subfamilies based on biological properties and nucleotide sequence comparison that are termed α, β, and γ. The α herpesviruses include herpes simplex virus 1 and 2, varicella zoster virus, and equine herpesvirus type 1; these viruses are neurotropic. The β herpesviruses include human cytomegalovirus (HCMV), murine cytomegalovirus (MCMV), and human herpesvirus types 6 and 7 (HHV-6, HHV-7); these viruses infect multiple cell types. The γ herpesviruses include Epstein–Barr virus (EBV), herpesvirus Saimiri (HVS), human herpes virus 8 or Kaposi's sarcoma-associated virus (KSHV), and equine herpesvirus type 2 (EHV-2); these viruses are tropic for lymphoid and epithelial cells and characteristically are capable of inducing transformation. A key feature of the herpesvirus family is the ability to establish latent or persistent infection of the host with the initial infection often being asymptomatic. Control of these asymptomatic infections relies upon an active immune response because, in the immunocompromised host, infection can be associated with considerable morbitity or mortality. In order to evade detection by the host immune response, herpesviruses are particularly adept at interfering with the recognition of infected cells by T cells via major histocompatibility complex (MHC) class I antigens (reviewed in ref. 4). However, herpesviruses also encode many proteins related to host cytokines and cytokine receptors that are used to enhance viral replication and evade immune detection.

3. Cytokines encoded by viruses

Computer-assisted analysis of herpes and poxvirus genomes has revealed that these viruses encode their own versions of many cytokines (Table 1). These may function

Table 1 Cytokines encoded by viruses

Cytokine	Virus	ORF	Receptor	Activity	Reference
Epidermal growth factor (EGF)	Vaccinia/variola/cowpox	C11R/D4R/C5R	EGFR	Agonist	5–7
	Myxoma/shope fibroma		EGFR	Agonist	9
Vascular endothelial growth factor (VEGF)	Orf		VEGFR?	Agonist ?	13
CC chemokine	Molluscum contagiosum	MC148R	CCRs/CXCRs	Antagonist	14–16
	Human herpes virus 8	K4/K6	CCRs/CXCRs	Antagonist/agonist	18–20
	Human herpes virus 6A	U83	?	?	22
	Murine CMV	M131	?	?	23
CXC chemokine	Marek's disease virus	Eco Q	CXCR5?	Agonist?	21
Interleukin 6 (IL-6)	Human herpes virus 8	K2	IL-6R	Agonist	18, 24, 25
Interleukin 10 (IL-10)	Epstein–Barr virus	BCRF-1	IL-10R	Agonist	29, 30
	Equine herpes virus 2		IL-10R?	?	35
	Orf		IL-10R?	Agonist	36
Interleukin 17 (IL-17)	Herpesvirus saimiri	13	IL-17R	Agonist	42

as growth factors for cells, facilitating the replication of the virus when these cells are infected, or may suppress immune responses at the site of infection. Further, some viruses have modified the mammalian cytokine during their evolution such that it antagonizes the activity of the cytokine to which it is related.

3.1 Viral growth factors

3.1.1 Poxvirus epidermal growth factor (EGF)

Genes encoding growth factors have been identified in the genomes of some poxviruses. A highly significant similarity was found between a 19 kDa VV protein and the epidermal growth factor (EGF) family (5–7). The VV protein, now called the vaccinia growth factor (VGF), is encoded by open reading frame (ORF) C11R in VV strain Copenhagen and has 45% amino acid identity to human EGF. The VGF is secreted from infected cells and binds to mammalian EGF receptors, thereby initiating signal transduction and a proliferative response. Related growth factor genes have been described in other poxviruses such as variola virus (8), and the *leporipoxviruses* such as Shope fibroma virus and myxoma virus, which infect rabbits (9). Deletion of the VGF gene from the VV genome reveals that the growth factor is non-essential for growth *in vitro*, but *in vivo* the protein plays an important role in the pathogenesis of infection (10). A VV strain that lacked VGF, replicated less well than parental virus within the brains of mice. Additionally, intradermal inoculation of the VGF deletion mutant virus into rabbits required 10- to 100-fold more virus than wild-type to produce a visible lesion (10). A significant difference between the lesions induced by the wild-type and VGF deletion mutant viruses was the amount of cellular proliferation or hyperplasia of the epidermal cells at the margin of the lesion. This suggested that the role of the VGF may be to stimulate hyperplasia at the site of infection and thereby increase virus replication in these actively dividing cells (11). In *leporipoxviruses*, deletion of the growth factor also has profound effects on viral pathogenesis. Whereas all animals infected with wild-type malignant rabbit fibroma virus (MRV) died, 75% of animals infected with growth factor deletion virus recovered. Once again, hyperplasia of epithelial cells was reduced in animals infected with the deletion mutant virus (12).

3.1.2 Poxvirus vascular endothelial growth factor (VEGF)

Sequencing of selected regions of orf virus (OV), a member of the *parapoxvirus* genus, which causes a highly contagious pustular dermatitis in sheep and goats and is transmissible to humans, revealed an ORF with amino acid similarity to vascular endothelial growth factor (VEGF) (13). VEGF is a heparin-binding, dimeric protein related to the platelet-derived growth factor (PDGF) family of growth factors. It is mitogenic for vascular endothelial cells and plays an important role in angiogenesis. Additionally, VEGF is chemoattractant for monocytes and enhances blood vessel permeability. *In vitro*, supernatants from cell cultures infected with OV have mitogenic effects on vascular endothelial cells, which may be caused by the viral VEGF. In addition, OV lesions are characterized by their highly vascularized nature, a feature that may be caused by the presence of the VEGF-like protein. How this is of advantage to the virus is unclear, but an increase in the flow of essential metabolites and indirect proliferation of epithelial tissue have been proposed as possible explanations (13).

3.2 Viral chemokines

3.2.1 Molluscum contagiosum virus (MCV) CC chemokine antagonist

Since the eradication of smallpox the only poxvirus to infect humans naturally is MCV. This virus produces small, papular benign tumours in the skin of healthy children and immunocompromised adults. Unlike the acute systemic disease caused by variola virus, MCV persists in the skin for months to years before regressing and can be a considerable problem in acquired immune deficiency syndrome (AIDS) patients, where extensive MCV infections can be disfiguring. Sequencing of the MCV genome (14) revealed a putative viral protein (MC148R) with significant amino acid similarity to CC chemokines. The N-terminus of some chemokines is responsible for inducing signal transduction following receptor engagement, since deletions in this region abrogate signalling. Comparison of the MC148R protein and known chemokines revealed a short deletion in the N-terminal region of the mature viral molecule. Therefore, the MC148R ORF was predicted to function as a chemokine receptor antagonist. This was confirmed in two studies. In chemotaxis assays the viral protein exhibited no chemotactic activity, but blocked the chemotactic response to human macrophage inflammatory protein-1α (MIP-1α) (15). In a more extensive study, the viral protein was shown to interfere with the chemotaxis of cells in response to both CC and CXC chemokines and to block calcium fluxes induced by the CC chemokines, monocyte chemotactic protein-1 (MCP-1) and I309, and the CXC chemokine, stromal cell-derived factor-1α (SDF-1α) (16). Thus, unlike the restricted specificity of host chemokines for one or a few receptors, the viral molecule appears to engage a large number of receptors (at least five host receptors). Obviously, this broad antagonistic activity could be of considerable advantage to the virus and may be the reason that MCV lesions are characterized by an absence of inflammatory cell infiltrate.

3.2.2 HHV-8 (Kaposi's sarcoma-associated herpesvirus) chemokines

Kaposi's sarcoma (KS) is a highly angiogenic neoplasm found most frequently in immunocompromised individuals (17). Increasing evidence implicates a new human herpesvirus (HHV-8) as the infectious agent responsible for this tumour. Like other γ herpesviruses, HHV-8 infects lymphocytes and hence infection has been associated with B cell lymphomas and the lymphoproliferative disorder, Castleman's disease.

Three chemokine genes are present within the HHV-8 genome. These are termed viral macrophage inflammatory proteins (vMIP-I, vMIP-II, and vMIP-III) (18). As yet little is known about vMIP-III except that it is distantly related to the family of CC chemokines. vMIP-I and vMIP-II share 48% amino acid identity and are more closely related to each other than to cellular chemokines, suggesting that they have arisen by a gene duplication event after the original gene was acquired from the host. Initially, vMIP-I was shown to block infection by CCR5 tropic HIV-1 strains, which suggests that vMIP-I bound CCR5 (18). Further studies have indicated that vMIP-II was a potent inhibitor of HIV infection of cells expressing CCR3, the eotaxin receptor that is expressed predominately on eosinophils. The binding of vMIP-II to CCR3 was confirmed in calcium-flux assays where the viral chemokine was a potent activator of human eosinophils and also induced eosinophil chemotaxis (19). Despite the similarities between vMIP-I and vMIP-II, vMIP-I was inactive on both eosinophils and neutrophils and neither viral chemokine demonstrated antagonistic activity. Certain chemokines have been shown to induce blood vessel formation and, therefore, it was of interest to examine the angiogenic potential of the viral MIPs. Both vMIP-I and vMIP-II induced angiogenesis on the chorioallantoic membrane of chick embryos and this may have important implications for the role of HHV-8 in KS lesions (19). The potent agonistic activity of vMIP-II for eosinophils contrasts with its antagonistic activity on other cells. In an independent study (20), vMIP-II was found to bind human chemokine receptors CCR1, CCR2, CCR5, and CXCR4, and blocked the calcium mobilization induced by relevant chemokines through these receptors and CCR3. Additionally, vMIP-II was a potent inhibitor of human monocyte chemotaxis in response to RANTES, MIP-1α, and MIP-1β. Thus, although vMIP-II has positive effects on angiogenesis and eosinophil recruitment, it also has negative effects on other chemokine functions. The blockade of HIV infection by the HHV-8 vMIPs hold out the hope that these proteins may serve as a guide for the development of broad spectrum anti-HIV agents.

3.2.3   Marek's disease virus (MDV) chemokine

Marek's disease virus is a lymphotropic avian α herpesvirus that causes lethal disease in chickens. The virus induces a lymphoproliferative disease with an early, cytolytic infection of B cells in the spleen, thymus, and bursa of Fabricius, and a subsequent induction of latent infection characterized by the generation of T cell lymphomas. After the cloning of the CXC chemokine, B lymphocyte chemoattractant (BLC), a molecule that plays an important role in the migration of B cells into lymphoid follicles, it was found that a related protein was encoded in the genome of MDV (21). The viral protein encoded by the Eco Q ORF exhibits a higher level of similarity to BLC than do other members of the human CXC chemokine family. BLC binds to the CXCR5 receptor and induces migration of B cells but not T cells, granulocytes, or macrophages. Although the biological activity of the viral molecule has not been reported it is possible that the MDV chemokine induces chemotaxis of B cells into the site of virus infection, thereby providing the virus with a rich source of its target cells.

3.2.4 β Herpes virus chemokines

Two other potential chemokines are encoded in herpesvirus genomes. Human herpesvirus 6 (HHV-6) is ubiquitous in the population and primarily infects CD4$^+$ T cells. HHV-6 consists of two variants, HHV-6A and HHV-6B. HHV-6B is the aeteological agent of 'roseola infantum' a childhood illness characterized by skin rash and high fever. HHV-6A has yet to be linked to any human disease but its genome has been sequenced (22). HHV-6A ORF U83 is predicted to encode a chemokine related to the CC family. Interestingly, while the N-terminal, receptor-binding region is conserved, the C-terminal region that is thought to interact with glycosaminoglycan presentation molecules, is highly divergent. However, the molecule has yet to be shown to be secreted or to have biological activity.

Murine cytomegalovirus (MCMV) ORF M131 also has similarities to CC chemokines (23), however, to date no biological activity has been described for this molecule.

3.3 HHV-8 IL-6

In addition to the HHV-8 encoded viral chemokine molecules (Section 3.2.2), the HHV-8 ORF K2 encodes an IL-6-like molecule (vIL-6) with 25% amino acid identity and 62% similarity to human IL-6 (18, 24, 25). Critically all four cysteines in human IL-6 are conserved in the viral molecule. The vIL-6 displays biological activity in IL-6 bioassays, inducing proliferation of the mouse plasmacytoma cell line, B9. *In vitro*, cells infected with HHV-8 produced vIL-6 constitutively and this was increased markedly by treatment with phorbol esters. Further studies have revealed that treatment of cells with vIL-6 leads to rapid induction of Janus kinase 1 (JAK1) phosphorylation and formation of a DNA binding complex containing signal transducer and activator of transcription (STAT)1 and STAT3. Although anti-IL-6 receptor α (IL-6Rα) chain antibodies blocked human IL-6-induced signal trans-duction, they failed to block vIL-6-induced responses. However, antibodies reactive with the gp130 component of the IL-6 receptor blocked responses caused by both human IL-6 and vIL-6. Interestingly, cells expressing the gp130 component of the IL-6 receptor alone respond to vIL-6 but not human IL-6, which indicates that vIL-6 can induce specific JAK/STAT signalling via gp130 independently of the IL-6Rα chain (26). In contrast, another study using antibodies against both the IL-6Rα chain and gp130 revealed that both receptor components are important in vIL-6 signalling (27).

The current role for vIL-6 in HHV-8 pathogenesis is unclear. However, given the biological activities of host IL-6, it is possible that vIL-6 may have roles in the prevention of virally-induced apoptosis and the induction of other cytokines such as VEGF, which leads to virally-induced angiogenesis. vIL-6 could also be responsible for the lymphoproliferative disorders associated with KS such as primary effusion lymphoma (PEL) and KS-associated Castleman's disease.

3.4 Viral IL-10 (vIL-10)

3.4.1 EBV IL-10

EBV is a ubiquitous γ herpesvirus and is the causative agent of infectious mono-nucleosis as well as being associated with Burkitt's lymphoma, Hodgkin's lymphoma, and nasopharyngeal carcinoma.

IL-10 is a cytokine produced by murine CD4$^+$ Th2 clones, B cells, keratinocytes, activated mast cells, and macrophages, and was originally described as cytokine synthesis inhibitory factor (CSIF) (28). IL-10's ability to inhibit activation of cytokine synthesis, especially IFN-γ, makes this cytokine a potent inhibitor of macrophage, NK cell, and T cell function (see also Chapter 3). During the cloning of IL-10 it was found that both human and murine IL-10 exhibit a remarkably high degree of amino acid identity to the EBV ORF, BCRF1 (29). Human IL-10 and BCRF1 share 84% amino acid identity, which suggests that there has been considerable restraint on the viral molecule to preserve its structure or that it was acquired from the host recently. Like the cellular molecule, BCRF-1, or vIL-10, is a secreted 17 kDa protein and is expressed in the late phase of the lytic cycle but not in latently infected cells (30). However, despite the high level of similarity between human IL-10 and vIL-10 the molecules show differences in their biological activities. vIL-10 exhibits a 1000-fold reduced ability to inhibit IL-2 production from T cell clones and this is correlated with a 1000-fold lower affinity of vIL-10 for the human IL-10Rα chain (31). However, many other biological activities of human IL-10 are shared by vIL-10. In addition to the inhibition of IFN-γ production, vIL-10 inhibits the production of superoxide anion (O_2-) from monocytes in response to various cytokines (32). Inhibition of IL-4-induced IgE synthesis via inhibition of monocyte accessory cell function has also been demonstrated (33). More recently, down-regulation of the peptide transporter TAP1 but not TAP2 by vIL-10 has been shown in B cells. This down-regulation impedes peptide transport into the endoplasmic reticulum and hence leads to a reduction in cell surface MHC class I antigen (34). This may be important in reducing the recognition of EBV-infected cells by CD8$^+$ cytotoxic cells during the lytic phase of replication when both early and late viral proteins are being produced by the infected cell.

3.4.2 Equine herpesvirus type 2 (EHV-2) IL-10

A gene predicted to encode an IL-10-like protein has also been found in the genome of EHV-2 (35), a slow growing cytopathogenic γ herpesvirus that is ubiquitous within the equine population. The ORF, present in the 4.3 kb DNA *Eco*RI N fragment, encodes a protein with 76% amino acid identity to human IL-10, which suggests that the molecule will have IL-10 activity, although this remains to be demonstrated.

3.4.3 Orf virus IL-10

In addition to the finding that OV encodes a VEGF-related protein (Section 3.1.2), sequence analysis of a 6.0 kb DNA fragment derived from the right terminus of the OV genome revealed an ORF with a high degree of amino acid identity to mammalian IL-10 and the IL-10-like proteins of EBV and EHV-2 (36). The highest similarity was seen with ovine IL-10 (80% identity) suggesting that the viral gene has been captured from its host during virus evolution. OV IL-10 was expressed transiently from COS cells and supernatants from these cultures expressed an IL-10-like activity in a murine thymocyte assay. The role of the OV IL-10 protein in the pathogenesis of viral infection or suppression of immune responses has yet to be reported.

3.5 HVS IL-17

Infection by HVS is asymptomatic in its natural host the squirrel monkey. However, in other New World primates such as marmosets and cottontop tamarins, the virus induces an acute T cell lymphoma. HVS is also capable of transforming human and simian T cells in culture to IL-2-dependent growth. The HVS ORF 13 was found to have 57% amino acid identity to a rodent protein termed CTLA8, subsequently termed IL-17 (37). This cytokine is expressed preferentially by CD4/CD8⁻ T cells, a small subset of T cells with immunoregulatory functions (38). The biological activities of IL-17 include the induction of other cytokines such as IL-6, IL-8, and granulocyte colony-stimulating factor (G-CSF) from stromal cells (39, 40) and pro-inflammatory cytokines such as IL-1 and TNF-α from human macrophages (41). An IL-17–human IgG Fc fusion protein (IL-17.Fc) and HVS13.Fc fusion proteins were used to identify and clone the IL-17 receptor (IL-17R) from the murine thymoma, EL-4 (42). The IL-17R cDNA encodes a type I membrane protein that is unrelated structurally to other cytokine receptor families and is widely distributed, in contrast to the ligand. IL-17 has been shown to support T cell proliferation raising the possibility that vIL-17 plays a role in virally-induced T cell transformation *in vivo*. However, compared with wild-type HVS, vIL-17 gene knock-out mutants show similar viral replication, T cell transformation *in vitro*, and virulence in cottontop tamarins (43). Therefore, the vIL-17 gene is not required for lymphoma induction but may play a role in viral persistence within the natural host. As both cellular and vIL-17 have been shown to induce activation of nuclear factor-κB (NF-κB), and induce production of cytokines such as IL-6, IL-8 (44), and G-CSF, expression of vIL-17 may activate surrounding cells, possibly indirectly via the viral IL-8 receptor (Section 4.2.5), allowing more efficient virus infection or replication.

4. Cytokine receptors encoded by viruses

4.1 Soluble cytokine receptors

Essentially all cellular cytokine receptors with a single transmembrane domain exist in a soluble form. Two mechanisms are known to result in the generation of soluble forms of cytokine receptors:

(a) Differential splicing of the primary mRNA transcript such that the sequences encoding the hydrophobic transmembrane domain are removed.

(b) Limited proteolytic cleavage in the extracellular domain of the cell surface receptor, which results in shedding.

Although most soluble receptors are predicted to be inhibitory for their cognate cytokine, there are soluble cytokine receptors with potent agonistic activity, for example, the soluble IL-6R α chain (45). The coincident explosion of cytokine receptor cloning in the late 1980s and early 1990s and the sequencing of large regions of poxvirus genomes revealed that these viruses have modified versions of cellular cytokine receptors in their genomes (Table 2). Subsequently, other viral soluble receptors have been characterized by functional assays and expression cloning.

4.1.1 Soluble poxvirus TNF receptors (TNFRs)

Cloning of the cellular TNFRs and subsequent database analysis revealed related molecules in poxvirus genomes (46, 47). Shope fibroma virus (SFV) ORF T2 encodes a secreted 58 kDa protein that binds both human TNF-α and TNF-β (46). A similar ORF was found in myxoma virus (M-T2) (48). Members of the TNFR family are characterized by ligand-binding cysteine-rich domains (CRDs) and the highest amino acid similarity between the TNFRs and the SFV and myxoma T2 ORF was seen within the first three of the four CRDs. Over these regions, the amino acid identity between the M-T2 ORF and the p75 TNFR was 43%. Deletion of the M-T2 ORF from myxoma resulted in attenuation of the virus as judged by clinical signs of disease and mortality in rabbits (48).

M-T2 is secreted from virus-infected cells as a monomer and a disulfide-linked dimer, both of which bind to rabbit TNF-α with high affinity (K_d of 170 pM and 195 pM, respectively) (49). However, despite the similar affinities, the dimeric M-T2 was a considerably more potent TNF-α inhibitor, presumably because of more efficient inhibition of cellular receptor oligomerization.

Myxoma virus mutants with a disrupted T2 ORF induce rapid apoptosis when used to infect rabbit T cells. Surprisingly, addition of exogenous T2 protein failed to reverse this effect, which suggests that T2 protein may also have intracellular functions (50). Mutational analysis of the M-T2 ligand-binding domain revealed that the three N-terminal CRDs are required for TNF-α binding (51). Deletion of the C-terminal region resulted in retention of TNF-α binding activity although the protein was poorly secreted. Interestingly, M-T2 C-terminal deletions that removed all but the first two N-terminal CRDs retained their ability to inhibit apoptosis when used to infect rabbit T cells (52). The mechanism by which this truncated, intracellular M-T2 prevents apoptosis will require identification of its cellular target proteins.

Similar genes have been found in the orthopoxviruses, cowpox virus (CrmB) (53), variola virus (G4R/G2R), and VV (B28R/C22L). However, in VV strains Copenhagen, WR, and Lister the genes are inactive because of frame shift and nonsense mutations (54, 55) The CrmB ORF is transcribed early during infection and, by ligand blotting, was shown to bind both TNF-α and TNF-β (53).

A second gene predicted to encode a TNFR-like protein was identified in the genomes of cowpox virus (CrmC) (56) and VV (A53R) (47) but interestingly this was absent from variola virus strains Harvey (57), Bangladesh 1975, and India 1967. As for the VV B28R/C22L ORFs, the A53R ORF in VV strains WR and Copenhagen is inactive but is active in Lister (54, 55). Similar to the T2/CrmB/B28R TNFRs, this ORF has the highest similarity to Type II (75 kDa) human TNFR and is composed of four CRDs, although the fourth domain is truncated. However, unlike the CrmB ORF, cowpox virus CrmC is transcribed at late times during infection and binds and inhibits the biological activity of TNF-α but not TNF-β (56). Deletion of both the CrmB and CrmC ORFs has no significant effect on pock morphology in the chorio-

Table 2 Soluble cytokine receptors/binding proteins encoded by viruses

Soluble cytokine receptor/binding protein	Virus	ORF	Ligands	Activity	Reference
Tumour necrosis factor	Shope fibroma/myxoma	T2	TNF-α/TNF-β	Antagonist	46, 48
	Vaccinia/variola/cowpox	B28R/G2R/CrmB(H4R)	TNF-α/TNF-β	Antagonist	53, 54
	Vaccinia/cowpox	A53R/CrmC(A53R)	TNF-α	Antagonist	56
	Cowpox/ectromelia	CrmD	TNF-α/TNF-β	Antagonist	58
Interleukin 1	Vaccinia/cowpox	B15R/B14R	IL-1β	Antagonist	62, 63
Interferon γ	Myxoma	T7	Rabbit IFN-γ/chemokines	Antagonist	66
	Vaccinia/variola/cowpox	B8R/B8R/B7R	Multiple species IFN-γ	Antagonist	67, 69
	Swinepox/camelpox/ectromelia	C6L			
Interferon α/β	Vaccinia/variola/cowpox/ camelpox/ectromelia	B18R/B17R/B17R	Multiple species IFN-α/β and IFN-ω	Antagonist	75, 76
Chemokine	Myxoma	T7	Chemokines/rabbit IFN-γ	Antagonist?	81
CC chemokine	Vaccinia/variola/cowpox/ camelpox/racoonpox	B29R/G3R/D1L	CC chemokines	Antagonist	82–84
	Myxoma/shope fibroma	T1	CC chemokines	Antagonist	82
Colony-stimulating factor-1	Epstein–Barr virus	BARF-1	CSF-1α,β,γ	Antagonist	86
IL-2/IL-5/IFN-γ	Tanapox	?	IL-2/IL-5/IFN-γ	Antagonist	87

allantoic membrane of chick embryos (56). To date, no virulence studies have been reported in mammalian hosts. Interestingly, although the viral TNFR N-terminal CRDs show similarity to the cellular TNFRs the C-terminal 160 amino acids of CrmB show no similarity but are highly conserved in orthopoxviruses and may bind to as yet unknown proteins.

More recently, a third TNFR has been discovered in cowpox virus Brighton red strain (CrmD) (58). This ORF was interrupted by frame shift mutations in other strains of cowpox and absent in the other orthopoxvirus variola virus and VV. However, the ORF was present in several strains of ectromelia virus. Secreted CrmD is composed of four CRDs and is released as disulfide-linked complexes that bind both TNF-α and TNF-β. The CrmD ORF contains a 160 amino acid C-terminal region with 44% identity to the C-terminal region in CrmB. The expression of up to three different TNFRs by cowpox virus suggests that each molecule may modulate the antiviral effects of TNF by different but complementary mechanisms.

Further complexity has been added by the finding that the Lister, USSR, and Evans strains of VV express a TNF-α receptor at the cell surface as well as in the supernatant (55). The gene encoding this cell surface TNFR has not yet been identified. The cell surface TNFR might function to protect virus-infected cells from the cytotoxic effects of TNF.

4.1.2   Other viral TNF receptors

Lymphocystis disease virus (LCDV) belongs to the *Iridoviridae* family and is the causative agent of a disease reported to occur in many different species of fish. Fish infected with this virus develop clusters of hypertrophied fibroblasts that are encapsulated within a hyaline extracellular matrix and contain enormous numbers of virions. Sequencing of the viral double-stranded DNA genome revealed an ORF (167L) with significant amino acid similarity to members of the TNFR family (59). The molecule is predicted to contain one complete and one incomplete CRD domain and has no potential transmembrane domain, leading to the prediction that the molecule is secreted. The highest amino acid similarity of ORF 167L was observed with the OX40 molecule. The biological activities of the 167L molecule remain to be reported.

4.1.3   Soluble poxvirus IL-1β receptor

The B15R ORF of VV WR strain was identified initially as a potential IL-1 receptor based on its sequence similarity to the cellular Type I IL-1R (60). Subsequent cloning of the Type II IL-1R revealed a closer similarity of this receptor to the B15R ORF (61). Similar to the cellular IL-1Rs, B15R encodes a member of the immunoglobulin superfamily (IgSF) with an N-terminal signal peptide and three Ig domains. However, B15R lacks a potential transmembrane domain, suggesting that the molecule would be secreted. Demonstration of IL-1 binding activity in supernatants from VV-infected cells, which was absent in B15R deletion mutants, and in the supernatants of mammalian cells and baculovirus-infected cells expressing B15R revealed that the B15R molecule was a soluble vIL-1R (62, 63). Interestingly, the vIL-1R bound IL-1β with high affinity (K_d 234 pM) but failed to demonstrate any significant binding to IL-1α or the IL-1 receptor antagonist (IL-1ra) (62). This binding profile is similar to the soluble form of the cellular Type II IL-1R, suggesting that the Type II IL-1R was captured by the virus either to inhibit IL-1β specifically or to avoid inhibitory action on the other IL-1 inhibitor, the IL-1ra (64).

The role of the vIL-1R in the pathogenesis of viral infection was investigated using B15R deletion mutants. In an intracranial infection model, the B15R deletion mutant was attenuated, exhibiting an LD_{50} 285-fold higher than the wild-type virus (63). However, when the intranasal route was used for infection, a similar deletion mutant demonstrated increased virulence (62). It was proposed that this increased virulence resulted from excessive levels of systemic host IL-1β and that, in this model, the vIL-1R abrogates the detrimental effects of virus-induced IL-1β on the host, hence promoting host survival and virus spread. The modulation of the host systemic response to viral infection by the vIL-1R was investigated further by the measurement of the febrile response (65). Mice infected with wild-type virus failed to develop fever, despite developing a severe infection, whereas those infected with the B15R deletion mutant developed a sustained fever from two to five days post-infection. Additionally, VV strains such as Copenhagen that do not express vIL-1R induced a febrile response that was abrogated in the case of VV strain Copenhagen by repair of the vIL-1R ORF (65). Interestingly, all variola virus strains sequenced to date are predicted not to express the vIL-1R because of frame shifts and nonsense mutations in the ORF. Therefore, it is relevant that smallpox was characterized by a high and sustained fever. These results demonstrate the pivotal role of IL-1β as an endogenous pyrogen in poxvirus infection and represent the first example of a virus controlling the host febrile response.

4.1.4 Soluble poxvirus IFN-γ receptor (IFN-γR)

A poxvirus IFN-γR was identified first in the supernatants of cells infected with myxoma virus (66). Peptide sequencing of a major secreted 37 kDa myxoma virus protein identified this as the product of the M-T7 ORF. Computational database searching revealed a low but significant similarity of M-T7 to the extracellular domains of human and mouse IFN-γR α chains. Cross-linking of radiolabelled IFN-γ to supernatants from myxoma-infected cells revealed a 51 kDa complex demonstrating that M-T7 was a functional vIFN-γR. Additionally, M-T7 containing supernatants effectively inhibited the antiviral activities of rabbit IFN-γ *in vitro*. Related proteins are expressed by a number of orthopoxviruses, with the VV protein encoded by the B8R ORF (67). Within VV strains, the vIFN-γR is highly conserved and, of those poxviruses examined, only two, modified virus Ankara (MVA), a highly attenuated strain of VV (68) and MCV (14), do not encode a vIFN-γR.

Mammalian IFN-γ exhibits a high degree of species specificity and, therefore, it was of interest to examine the species specificity of the poxvirus vIFN-γR. The myxoma virus vIFN-γR demonstrated a strict species specificity for rabbit IFN-γ, which is consistent with its restricted host tropism. However, vIFN-γRs from orthopoxviruses including VV, cowpox, and camelpox bound a wide range of IFN-γs including human, cow, rat, and rabbit, but not mouse IFN-γ (67, 69). The binding of the VV vIFN-γR to rat but not mouse IFN-γ was unexpected given the high degree of amino acid identity (87%) between the two molecules. Surprisingly, the VV vIFN-γR also binds to chicken IFN-γ (70). The IFN-γR from ectromelia virus, the causative agent of mousepox, also has a broad species specificity but this includes binding to mouse IFN-γ (69). The broad species specificity of the orthopoxvirus vIFN-γR may have aided virus replication in multiple host species during virus evolution.

Given the important role that IFN-γ plays in restriction of poxvirus replication in the host (71, 72) it would be expected that the vIFN-γR is a critical virulence factor for poxvirus infection. Recently this was demonstrated in myxoma virus; deletion of the M-T7 ORF from myxoma virus resulted in a dramatic reduction in signs of disease and viral dissemination when compared with the wild-type virus (73). Additionally, wild-type viral lesions contained few infiltrating leukocytes whereas M-T7 deletion mutant lesions were characterized by extensive inflammatory infiltrates. However, interpretation of this data is complicated by the finding that in addition to binding IFN-γ, the M-T7 molecule also binds to a broad range of chemokines (Section 4.1.6).

4.1.5 Soluble poxvirus IFN-α/β receptor (IFN-α/βR)

In addition to the VV B15R ORF (Section 4.1.3) another ORF (B18R) near the right-hand inverted terminal repeat of VV WR was identified as a member of the IgSF (60). This ORF shares 20% amino acid identity with B15R, but did not encode an IL-1 binding protein (62). Like B15R, the B18R ORF was predicted to encode a soluble protein because of the lack of a C-terminal hydrophobic membrane anchor. However, it was known that B18R protein associated with the cell surface as it had been identified independently as the poxvirus S antigen (74). Surprisingly, the B18R ORF was shown to encode an IFN-α/β binding and inhibitory activity (75, 76). Although a limited sequence similarity between B18R and the cellular IFN-α/βR was reported (76), the finding that a member of the IgSF bound to IFN-α/β was unexpected given that the cellular IFN-α/βRs belong to the class II cytokine receptor family and are composed of fibronectin type III domains.

The vIFN-α/βR binds to all human Type I IFNs tested to date (75–77), but not to IFN-γ, and demonstrates high affinity binding to human IFN-α2 (K_d 174 pM). Like the orthopoxvirus vIFN-γR, the vIFN-α/βR bound to Type I IFNs from several different species including man, rabbit, cow, rat, and pig but exhibited low affinity binding to mouse Type I IFN (75, 77). The binding of the vIFN-α/βR to human IFN-α1 and IFN-α2 in comparison to the cellular receptor was investigated using mono-clonal antibodies that recognize distinct regions of IFN-α1 and α2. Only antibodies against the N-terminus of IFN-α1 and IFN-α2 blocked binding to the cellular IFN-α/βR, whereas antibodies against both the N- and C-termini were unable to bind to IFNs that were bound to the vIFN-α/βR (78). This data indicate a more intimate interaction between IFN-α and the vIFN-α/βR than that between IFN-α and the cellular IFN-α/βR. This more intimate interaction may generate a more efficient Type I IFN inhibitor and allow the vIFN-α/βR to bind a larger number of Type I IFN species.

The vIFN-α/βR is conserved in 14 strains of VV and in cowpox and camelpox viruses. Interestingly, two strains of VV, Lister and MVA, failed to produce a vIFN-α/βR, and VV strain Wyeth produced a vIFN-α/βR with a 78-fold lower affinity (68, 75). Use of these three vaccine strains was associated with the lowest level of post-vaccinial neurological complications associated with smallpox vaccination; possibly this is in part because of the lack of a functional vIFN-α/βR. The role of the vIFN-α/βR in virus virulence was investigated using a vIFN-α/βR deletion mutant in a mouse model. Deletion of the vIFN-α/βR attenuated the virus when compared with wild-type virus when administered via the intranasal route, and diminished viral replication in the lungs and dissemination to the brain (75).

In addition to acting as a soluble vIFN-α/βR, cell associated vIFN-α/βR was able to bind Type I IFN and act as a decoy receptor blocking Type I IFN-induced signal transduction (76). Presumably, retention of the vIFN-α/βR on the cell surface generates a microenvironment where the antiviral effects of Type I IFN on uninfected cells are abrogated to provide a population of cells permissive for subsequent virus infection.

4.1.6 Soluble poxvirus chemokine binding proteins

To date, all cellular chemokine receptors (CKRs) have been found to belong to the serpentine family of seven transmembrane G protein-coupled proteins. Poxvirus ORFs with similarity to these proteins have been found in swinepox and capripox viruses (79, 80) (Table 3). Computational analysis of orthopoxvirus genomes revealed no ORFs with similarity to cellular CKRs. However, three groups reported independently that soluble chemokine binding proteins (CKBPs) were released from *orthopoxvirus-* and *leporipoxvirus*-infected cells.

In addition to binding rabbit IFN-γ, the M-T7 protein (Section 4.1.4) was also found to bind to CXC, CC, and C chemokines (81). Binding of radiolabelled RANTES to MT-7 was competed with both RANTES and rabbit IFN-γ. Interestingly, N-terminal chemokine deletion mutants bound normally to MT-7, but C-terminal chemokine deletion mutants that had lost the heparin binding domain failed to bind. Additionally, heparin competed the binding of chemokines but not IFN-γ to MT-7. As chemokines interact with their cognate receptors via the N-terminus it is likely that MT-7 will not prevent binding of chemokines to CKRs on leukocytes. However, as chemokines are thought to form gradients via their C-terminus on cell surface proteoglycans, MT-7 may disrupt chemokine localization or presentation to infiltrating leukocytes.

A second poxvirus CKBP was discovered by cross-linking of radiolabelled chemokines to supernatants from poxvirus-infected cells (82, 83). Subsequently, the activity encoded by myxoma virus was mapped to the T1 gene (82) and the activity from orthopoxviruses was mapped to a gene encoding a 35 kDa protein (82, 83). Simultaneously, the 35 kDa protein was found to bind the CC chemokine, MCP-1, by 'ligand fishing' using a biosensor-based analytical system (84). The viral CKBP (vCKBP) is expressed by many poxviruses including vaccinia, cowpox, camelpox, variola, Shope fibroma, myxoma, and racoonpox. Chemokine binding studies re-

Table 3 Chemokine receptor-like proteins encoded by viruses

Virus	ORF	Ligands	Similarity/function	Reference
β herpesvirus				
Human herpes virus 6A	U12	RANTES/MIP-1α/ MIP-1β/MCP-1	Related to UL33, M33	97
	U51		Similarity to opioid receptors	22
Human herpes virus 7	U12		Related to UL33, M33	
	U51		Similarity to opioid receptors	
Human CMV	US28	RANTES/MIP-1α/ MCP-1	Consumption of CC chemokines	90–94
	UL33		Related to UL12, M33/virion protein	102
	US27		Related to US28	
	UL78		Similarity to fMLP receptor	
Murine CMV	M33		Increases viral replication/dissemination	95
γ herpesvirus				
Human herpes virus 8	74	IL-8/NAP-2/GROα	Constitutively active/transforming?	98
Herpesvirus saimiri	74/ECRF3	IL-8/NAP-2/GROα	Related to HHV-8 ORF 74	100
Murine γ herpesvirus 68	74		Related to HHV-8/HVS ORF 74	103
Equine herpesvirus 2	74		Related to HHV-8/HVS ORF 74	104
	E1		Related to human CCR1	
	E6		Unknown	
Poxviruses				
Swinepox	K2R		Similarity to human IL-8 chemokine receptor	79
Sheep-pox	Q2/3L		Similarity to human CC chemokine receptors	80

vealed that the vCKBP binds preferentially to the CC family of chemokines. The VV vCKBP bound to CC chemokines with high affinity, e.g. human MIP-1α ($K_d = 103 \pm 4$ pM) this is at least tenfold higher than the affinity that CC chemokines display for their cellular CKRs and suggests that the vCKBP would be a potent inhibitor of CC chemokine activity (83, 84). Consistent with this, the vCKBP blocked the binding of CC but not CXC chemokines to their cell surface CKRs, and inhibited the biological activity of CC chemokines as demonstrated by the inhibition of intracellular calcium mobilization and cellular migration *in vitro* (83, 84). In contrast to the MT-7 IFN-γ/CKBP, heparin fails to compete the binding of CC chemokines to the vCKBP (83), suggesting that the vCKBP interacts with chemokines through their N-terminal receptor-binding domain. The conservation of vCKBP in many different poxviruses and its high affinity for CC chemokines suggests that this protein may play an important role in the inhibition of inflammatory cell infiltration into the site of poxvirus infection. An initial study using a rabbitpox (a strain of VV) deletion mutant lacking the vCKBP gene reported similar virulence to wild-type virus in both mice and rabbits (85). However, subsequently the same rabbitpox deletion mutant used at higher doses was reported to cause a greater migration of inflammatory cells into dermal sites of infection in comparison to the wild-type virus (82). Further studies will be required to resolve this ambiguous data and should include virus mutants made with DNA from the homologous virus and include revertant viruses.

The inhibition of chemokine action represents an attractive therapeutic target. However, the multiple transmembrane structure of the cellular chemokine receptors makes the construction of a soluble version extremely difficult. To date, the most promising chemokine antagonists developed for clinical use have been N-terminal chemokine mutants that may act as receptor antagonists. The finding of a poxvirus, MCV (Section 3.2.1), with such a molecule and a herpesvirus, HHV-8 (Section 3.2.2), with a partial chemokine antagonist gives credibility to such an approach. However, the use of a soluble CKBP may also be another useful strategy. The finding that the vCKBP when injected intradermally could block the migration of eosinophils in response to the CC chemokine, eotaxin, emphasizes the potential of it, or its derivatives, as a therapeutic agent in inflammatory and allergic diseases (83). The origin of the poxvirus vCKBP remains enigmatic because to date no cellular protein with amino acid similarity has been described.

4.1.7 Soluble EBV colony-stimulating factor-1 (CSF-1) receptor

Until recently viral soluble cytokine receptors have been found only in the supernatants from poxvirus-infected cells. However, the finding that EBV produces a soluble colony-stimulating factor-1 receptor (CSF-1R), encoded by the BARF-1 ORF, now extends this mechanism of immune evasion to the herpesvirus family (86). Computer-assisted analysis initially revealed that the BARF-1 ORF could encode a secreted protein, having an N-terminal hydrophobic signal sequence and no transmembrane domain. Subsequently, a BARF-1.Fc fusion protein was used to identify a ligand on activated peripheral blood T cells and, using an expression cloning strategy, this ligand was demonstrated to be the β form of CSF-1. Further experiments revealed that the BARF-1 protein bound to all forms (α, β, and γ) of CSF-1 and was able to neutalize the biological activity of human CSF-1 in bone marrow proliferation assays. The EBV BARF-1 protein shows very limited similarity to the cellular CSF-1 receptors. However, a block of 13 amino acids in the BARF-1 molecule is conserved in the extracellular domain of members of the tyrosine kinase receptor family, which include the human CSF-1R, and suggest that the BARF-1 ORF originated as a cellular gene. Relatively little is known about the biological role of CSF-1 apart from its action as a macrophage growth and differentiation factor. Anti-CSF-1 antibodies can block antigen-driven T cell proliferation and CSF-1 protein enhances complement component C3, integrin, and cytokine expression from monocytes. The expression of a soluble CSF-1 receptor by a virus suggests an as yet unrecognized role for CSF-1 in antiviral defence, especially against EBV.

4.1.8 Tanapox virus IL-2/IL-5/IFN-γ binding protein

Tanapox virus is a member of the *Yatapoxvirus* genus and produces a mild disease in humans that is characterized by transient fever, nodular skin lesions, and local lymphoadenopathy. The virus produces a soluble 38 kDa glycoprotein that binds to human IL-2, IL-5, and IFN-γ but not to other cytokines (87). The 38 kDa protein inhibited the biological activity of all three cytokines. However, the ORF encoding this potentially remarkable molecule remains to be isolated and expressed.

4.2 Viral membrane CKRs

A large number of potential seven transmembrane CKR have been acquired by large DNA viruses (Table 3). Many of these ORFs remain uncharacterized and this review is restricted to those where chemokine binding or functional activity has been described.

4.2.1 HCMV chemokine receptors (US28)

Of the four potential seven transmembrane G protein-coupled receptors encoded within the HCMV genome, US28 has been the most extensively characterized (88, 89). The ORF has been shown to encode a calcium-mobilizing receptor for the CC chemokines RANTES, MIP-1α, and MCP-1, but not the CXC chemokines IL-8 or IFN-γ-inducible protein 10 (IP-10) (90). The binding of CC chemokines to the US28 protein has been shown to occur with high affinity: MCP-1, K_d = 600 pM; RANTES, K_d = 270 pM; MIP-1α, K_d = 1.2 nM; and MIP-1β, K_d = 7.5 nM (91). Infection of fibroblasts with HCMV results in the production of RANTES within eight hours of infection. However, by 48 hours post-infection RANTES was undetectable in the supernatant (92). This sequestration of extracellular RANTES has been shown to be mediated by the US28 receptor because deletion of US28, or both US28 and US27, fail to down-regulate RANTES. In contrast, cells infected with the US27 deletion mutant continue to down-regulate RANTES (93, 94). The capacity of the US28 CKR to sequester CC chemokines could abrogate local immune responses thereby preventing inflammatory cell recruitment to the site of viral infection.

4.2.2 MCMV chemokine receptor (M33)

The M33 ORF (95) is colinear with the HCMV UL33 gene and the predicted proteins share 47% amino acid identity. Similar ORFs are also found in HHV-6 (U12) (22) and HHV-7 (U12) (96). Transcription of M33 occurs early and late after infection. The M33 ORF was disrupted by insertion of the beta-galactosidase gene (95), but the growth of this mutant virus *in vitro* was indistinguishable from the wild-type MCMV. However, when used to infect mice, the M33 mutant virus demonstrated a severely restricted growth in the salivary glands when compared with wild-type virus (95). This is the first evidence that a viral CKR may play an important role in virus dissemination or replication *in vivo*.

4.2.3 HHV-6A CKR (U12)

The HHV-6A ORF U12 encodes a G protein-coupled receptor related to HCMV UL33 and MCMV M33 ORFs that is expressed late during infection. Recently, the U12 gene has been cloned and stably expressed in human erythroleukaemia cells. The receptor induces mobilization of intracellular calcium in response to the CC chemokines RANTES, MIP-1α, MIP-1β, and MCP-1, but not the CXC chemokine IL-8 (97).

4.2.4   HHV-8 chemokine receptor (ORF 74)

Like HVS, another γ herpesvirus, HHV-8 encodes a CKR (ORF 74). The binding characteristics of the receptor were characterized with labelled IL-8 and cross competition
with other chemokines. The results revealed that the HHV-8 CKR bound to CXC
chemokines but with relatively low affinity: IL-8, K_d = 25 nM; neutrophil activating
protein-2 (NAP-2), K_d = 23 nM; growth-related oncogene α (GROα), K_d = 270 nM.
The receptor also bound to CC chemokines but with K_d values greater than 100 nM,
which are unlikely to be significant biologically. Expression of the HHV-8 CKR in
COS cells resulted in constitutive (agonist-independent) signalling activity, raising
the possibility that this receptor was involved in the cellular transformation properties associated with HHV-8 infection. This was confirmed in a further study where
signalling through the HHV-8 CKR led to cellular transformation and tumourigenicity in nude mice (98). In addition, HHV-8 CKR expression led to endothelial cell
growth and the expression of VEGF *in vitro*, which could be responsible for the KS-
associated angiogenic and proliferative response. Interestingly, recently it has been
found that the non-ELR CXC chemokine IP-10, but not the related molecule, mono-
kine induced by IFN-γ (MIG), acts as an antagonist of the HHV-8 CKR, converting it
from an active to an inactive state (99). This finding will allow the hypothesis that
HHV-8 CKR expression and signalling leads to tumourgenicity to be tested directly.

4.2.5   HVS CKR (ECRF3/ORF 74)

The CKR ORF within the HVS genome (ECRF3) is colinear with the equivalent CKR
in other γ herpesviruses, ORF 74. ECRF3 shows approximately 30% amino acid identity
to the cellular IL-8 receptors. Expression of ECRF3 in *Xenopus* oocytes and subsequent calcium mobilization studies revealed that the ECRF3 receptor bound to the
human CXC chemokines IL-8, GROα, and NAP-2, however, no binding to human
CC chemokines was detected (100). Whether, like the HHV-8 CKR, the ECRF3 ORF
plays a role in the transforming activity of HVS remains to be tested.

5.   Conclusions and future prospects

Clearly viruses have exploited the host cytokine network to their advantage and the
current extensive study of viral immunomodulatory genes will expand our knowledge of the immune system in a number of ways. First, expression of a cytokine-
related protein by a virus provides important clues to the nature of the interaction
between the virus and its host. Secondly, the study of viral immunomodulatory
genes reveals new mechanisms of viral pathogenesis. Using this information it will
be possible to engineer attenuated but immunogenic live recombinant vaccines and
possibly target currently untreatable viral infections with novel antiviral strategies.
Thirdly, the study of viral immunomodulatory genes can guide us to the best therapeutic regimens for the treatment of inflammatory disease (for further discussion of
this see Chapters 3 and 8). A good example of this is the poxvirus soluble IL-1βR
(Section 4.1.3). This virus protein is similar in function to the soluble form of the

human Type II IL-1R (sIL-1R II). These proteins share a high binding affinity for IL-1β and a low affinity for IL-1α and the other IL-1 inhibitor, the IL-1ra. These binding properties allow the inhibition of the agonist without interfering with the natural antagonist. Clinical trials with an engineered soluble version of the Type I IL-1R (sIL-1R I), which binds equally well to all forms of IL-1, showed that the sIL-1R I had no anti-inflammatory effect following endotoxin administration to human volunteers because of its neutralization of IL-1ra inhibitory function (101). Use of the sIL-1R II, the receptor chosen by the virus, may prove more efficacious.

To date, most of the cytokine-related proteins in viruses represent ORFs that were identified because of sequence similarity to cellular proteins that had been previously isolated and characterized biologically. However, already a few examples exist of viral molecules that may represent novel cellular molecules. Probably the best example of this is the poxvirus vCKBP (Section 4.1.6). This molecule is as yet not represented in the mammalian protein databases. Certainly, identification of a cellular counterpart of the vCKBP would provide an extremely useful therapeutic agent for the treatment of inflammatory and allergic disease. It is of course possible that poxviruses have developed this molecule during their own evolution, nevertheless, a structural understanding of the interaction between the vCKBP and its ligands may provide the basis for the rationale design of chemokine/CKR antagonists.

Viruses undoubtedly contain many more immunoregulatory genes yet to be discovered. The characterization of these novel viral genes will lead to the identification of novel cellular genes or the functional characterization of cDNAs currently residing in EST databases.

References

1. Moss, B. (1996) Poxviridae: the viruses and their replication. In *Fields virology* (ed. B. N. Fields, D. M. Knipe, and P. M. Howley), p. 2637. Lippencott Raven Press, New York.
2. Smith, G. L., Symons, J. A., Khanna, A., Vanderplasschen, A., and Alcamí, A. (1997) Vaccinia virus immune evasion. *Immunol. Rev.,* **159**, 137.
3. McGeoch, D. J., Cook, S., Dolan, A., Jamieson, F. E., and Telford, E. A. (1996) Molecular phylogeny and evolutionary timescale for the family of mammalian herpesviruses. *J. Mol. Biol.,* **247**, 443.
4. Wiertz, E. J., Mukherjee, S., and Ploegh, H. L. (1997) Viruses use stealth technology to escape from the host immune system. *Mol. Med. Today,* **3**, 116.
5. Blomquist, M. C., Hunt, L. T., and Barker, W. C. (1984) Vaccinia virus 19-kilodalton protein: relationship to several mammalian proteins, including two growth factors. *Proc. Natl. Acad. Sci. USA,* **81**, 7363.
6. Reisner, A. H. (1985) Similarity between the vaccinia virus 19K early protein and epidermal growth factor. *Nature,* **313**, 801.
7. Brown, J. P., Twardzik, D. R., Marquardt, H., and Todaro, G. J. (1985) Vaccinia virus encodes a polypeptide homologous to epidermal growth factor and transforming growth factor. *Nature,* **313**, 491.
8. Massung, R. F., Esposito, J. J., Liu, L., *et al.* (1993) Potential virulence determinants in terminal regions of variola smallpox virus genome. *Nature,* **366**, 748.

9. Chang, W., Upton, C., Hu, S. L., Purchio, A. F., and McFadden, G. (1987) The genome of Shope fibroma virus, a tumorigenic poxvirus, contains a growth factor gene with sequence similarity to those encoding epidermal growth factor and transforming growth factor alpha. *Mol. Cell. Biol.*, **7**, 535.

10. Buller, R. M., Chakrabarti, S., Cooper, J. A., Twardzik, D. R., and Moss, B. (1988) Deletion of the vaccinia virus growth factor gene reduces virus virulence. *J. Virol.*, **62**, 866.

11. Buller, R. M., Chakrabarti, S., Moss, B., and Fredrickson, T. (1988) Cell proliferative response to vaccinia virus is mediated by VGF. *Virology*, **164**, 182.

12. Opgenorth, A., Strayer, D., Upton, C., and McFadden, G. (1992) Deletion of the growth factor gene related to EGF and TGF alpha reduces virulence of malignant rabbit fibroma virus. *Virology*, **186**, 175.

13. Lyttle, D. J., Fraser, K. M., Fleming, S. B., Mercer, A. A., and Robinson, A. J. (1994) Homologs of vascular endothelial growth factor are encoded by the poxvirus Orf virus. *J. Virol.*, **68**, 84.

14. Senkevich, T. G., Bugert, J. J., Sisler, J. R., Koonin, E. V., Darai, G., and Moss, B. (1996) Genome sequence of a tumorigenic poxvirus: prediction of specific host response–evasion genes. *Science*, **273**, 813.

15. Krathwohl, M. D., Hromas, R., Brown, D. R., and Broxmeyer, H. E. (1997) Functional characterization of the C-C chemokine-like molecules encoded by molluscum contagiosum virus Types 1 and 2. *Proc. Natl. Acad. Sci. USA*, **94**, 9875.

16. Damon, I., Murphy, P. M., and Moss, B. (1998) Broad spectrum chemokine antagonistic activity of a human poxvirus chemokine homolog. *Proc. Natl. Acad. Sci. USA*, **95**, 6403.

17. Boshoff, C. and Weiss, R. A. (1997) Aetiology of Kaposi's sarcoma: current understanding and implications for therapy. *Mol. Med. Today*, **3**, 488.

18. Moore, P. S., Boshoff, C., Weiss, R. A., and Chang, Y. (1996) Molecular mimicry of human cytokine and cytokine response pathway genes by KSHV. *Science*, **274**, 1739.

19. Boshoff, C., Endo, Y., Collins, P. D., *et al.* (1997) Angiogenic and HIV-inhibitory functions of KSHV-encoded chemokines. *Science*, **278**, 290.

20. Kledal, T. N., Rosenkilde, M. M., Coulin, F., *et al.* (1997) A broad-spectrum chemokine antagonist encoded by Kaposi's sarcoma- associated herpesvirus. *Science*, **277**, 1656.

21. Gunn, M. D., Ngo, V. N., Ansel, K. M., Ekland, E. H., Cyster, J. G., and Williams, L. T. (1998) A B cell homing chemokine made in lymphoid follicles activates Burkitt's lymphoma receptor-1. *Nature*, **391**, 799.

22. Gompels, U. A., Nicholas, J., Lawrence, G., Jones, M., Thomson, B. J., Martin, M. E., *et al.* (1995) The DNA sequence of human herpesvirus-6: structure, coding content and genome evolution. *Virology*, **209**, 29.

23. MacDonald, M. R., Li, X., and Virgin, H. W. (1997) Late expression of a β chemokine homolog by murine cytomegalovirus. *J. Virol.*, **71**, 1671.

24. Neipel, F., Albrecht, J. C., Ensser, A., Huang, Y. Q., Li, J. J., Friedman-Kien, A. E., *et al.* (1997) Human herpesvirus 8 encodes a homolog of interleukin-6. *J. Virol.*, **71**, 839.

25. Nicholas, J., Ruvolo, V. R., Burns, W. H., *et al.* (1997) Kaposi's sarcoma-associated human herpesvirus-8 encodes homologues of macrophage inflammatory protein-1 and interleukin-6. *Nature Med.*, **3**, 287.

26. Molden, J., Chang, Y., You, Y., Moore, P. S., and Goldsmith, M. A. (1997) A Kaposi's sarcoma-associated herpesvirus-encoded cytokine homolog (vIL- 6) activates signaling through the shared gp130 receptor subunit. *J. Biol. Chem.*, **272**, 19625.

27. Burger, R., Neipel, F., Fleckenstein, B., Savino, R., Ciliberto, G., Kalden, J. R., *et al.* (1998) Human herpesvirus type 8 interleukin-6 homologue is functionally active on human myeloma cells. *Blood*, **91**, 1858.

28. Moore, K. W., O'Garra, A., de Waal Malefyt, R.,Vieira, P., and Mosmann, T. R. (1993) Interleukin-10. *Annu. Rev. Immunol.*, **11**, 165.

29. Moore, K. W., Vieira, P., Fiorentino, D. F., Trounstine, M. L., Khan, T. A., and Mosmann, T. R. (1990) Homology of cytokine synthesis inhibitory factor (IL-10) to the Epstein–Barr virus gene BCRF1. *Science*, **248**, 1230.

30. Hsu, D. H., de Waal Malefyt, R., Fiorentino, D. F., Dang, M. N., Vieira, P., de Vries, J., *et al.* (1990) Expression of interleukin-10 activity by Epstein–Barr virus protein BCRF1. *Science*, **250**, 830.

31. Ying, L., de Waal Malefyt, R., Briere, F., Parham, C., Bridon, J. M., Banchereau, J., *et al.* (1997) The EBV IL-10 homologue is a selective agonist with impaired binding to the IL-10 receptor. *J. Immunol.*, **158**, 604.

32. Niiro, H., Otsuka, T., Abe, M., Satoh, H., Ogo, T., Nakano, T., *et al.* (1992) Epstein–Barr virus BCRF1 gene product (viral interleukin 10) inhibits superoxide anion production by human monocytes. *Lymphokine Cytokine Res.*, **11**, 209.

33. Punnonen, J., de Waal Malefyt, R., van Vlasselaer, P., Gauchat, J. F., and de Vries, J. E. (1993) IL-10 and viral IL-10 prevent IL-4-induced IgE synthesis by inhibiting the accessory cell function of monocytes. *J. Immunol.*, **151**, 1280.

34. Zeidler, R., Eissner, G., Meissner, P., Uebel, S., Tampe, R., Lazis, S., *et al.* (1997) Downregulation of TAP1 in B lymphocytes by cellular and Epstein–Barr virus-encoded interleukin-10. *Blood*, **90**, 2390.

35. Rode, H. J., Janssen, W., Rosen-Wolff, A., Bugert, J. J., Thein, P., Becker, Y., *et al.* (1993) The genome of equine herpesvirus type 2 harbours an interleukin (IL-10)-like gene. *Virus Genes*, **7**, 111.

36. Fleming, S. B., McCaughan, C. A., Andrews, A. E., Nash, A. D., and Mercer, A. A. (1997) A homolog of interleukin-10 is encoded by the poxvirus orf virus. *J. Virol.*, **71**, 4857.

37. Rouvier, E., Luciani, M. F., Mattei, M. G., Denizot, F., and Golstein, P. (1993) CTLA-8, cloned from an activated T cell, bearing AU-rich messenger RNA instability sequences and homologous to a herpesvirus Saimiri gene. *J. Immunol.*, **150**, 5445.

38. Kennedy, J., Rossi, D. L., Zurawski, S. M., Vega, F. Jr., Kastelein, R. A., Wagner, J. L., *et al.* (1996) Mouse IL-17: a cytokine preferentially expressed by alpha beta TCR + CD4-CD8-T cells. *J. Interferon Cytokine Res.*, **16**, 611.

39. Cai, X. Y., Gommoll, C. P. Jr., Justice, L., Narula, S. K., and Fine, J. S. (1998) Regulation of granulocyte colony-stimulating factor gene expression by interleukin-17. *Immunol. Lett.*, **62**, 51.

40. Chabaud, M., Fossiez, F., Taupin, J. L., and Miossec, P. (1998) Enhancing effect of IL-17 on IL-1-induced IL-6 and leukemia inhibitory factor production by rheumatoid arthritis synoviocytes and its regulation by Th2 cytokines. *J. Immunol.*, **161**, 409.

41. Jovanovic, D. V., Di Battista, J. A., Martel-Pelletier, J., Jolicoeur, F. C., He, Y., Zhang, M., *et al.* (1998) IL-17 stimulates the production and expression of pro-inflammatory cytokines, IL-beta and TNF-alpha, by human macrophages. *J. Immunol.*, **160**, 3513.

42. Yao, Z., Fanslow, W. C., Seldin, M. F., Rousseau, A. M., Painter, S. L., Comeau, M. R., *et al.* (1995) Herpesvirus Saimiri encodes a new cytokine, IL-17, which binds to a novel cytokine receptor. *Immunity*, **3**, 811.

43. Knappe, A., Hiller, C., Niphuis, H., *et al.* (1998) The interleukin-17 gene of herpesvirus Saimiri. *J. Virol.*, **72**, 5797.

44. Yao, Z., Painter, S. L., Fanslow, W. C., Ulrich, D., Macduff, B. M., Spriggs, M. K., *et al.* (1995) Human IL-17: a novel cytokine derived from T cells. *J. Immunol.*, **155**, 5483.

45. Rose-John, S. and Heinrich, P. C. (1994) Soluble receptors for cytokines and growth factors: generation and biological function. *Biochem. J.*, **300**, 281.

46. Smith, C. A., Davis, T., Wignall, J. M., Din, W. S., Farrah, T., Upton, C., *et al.* (1991) T2 open reading frame from Shope fibroma virus encodes a soluble form of the TNF receptor. *Biochem. Biophys. Res. Commun.*, **176**, 335.

47. Howard, S. T., Chan, Y. S., and Smith, G. L. (1991) Vaccinia virus homologues of the Shope fibroma virus inverted terminal repeat proteins and a discontinuous ORF related to the tumor necrosis factor receptor family. *Virology*, **180**, 633.

48. Upton, C., Macen, J. L., Schreiber, M., and McFadden, G. (1991) Myxoma virus expresses a secreted protein with homology to the tumor necrosis factor receptor gene family that contributes to viral virulence. *Virology*, **184**, 370.

49. Schreiber, M., Rajarathnam, K., and McFadden, G. (1996) Myxoma virus T2 protein, a tumor necrosis factor (TNF) receptor homolog, is secreted as a monomer and dimer that each bind rabbit TNFalpha, but the dimer is a more potent TNF inhibitor. *J. Biol. Chem.*, **271**, 13333.

50. Macen, J. L., Graham, K. A., Lee, S. F., Schreiber, M., Boshkov, L. K., and McFadden, G. (1996) Expression of the myxoma virus tumor necrosis factor receptor homologue and M11L genes is required to prevent virus-induced apoptosis in infected rabbit T lymphocytes. *Virology*, **218**, 232.

51. Schreiber, M. and McFadden, G. (1996) Mutational analysis of the ligand-binding domain of M-T2 protein, the tumor necrosis factor receptor homologue of myxoma virus. *J. Immunol.*, **157**, 4486.

52. Schreiber, M., Sedger, L., and McFadden, G. (1997) Distinct domains of M-T2, the myxoma virus tumor necrosis factor (TNF) receptor homolog, mediate extracellular TNF binding and intracellular apoptosis inhibition. *J. Virol.*, **71**, 2171.

53. Hu, F.-Q., Smith, C. A., and Pickup, D. J. (1994) Cowpox virus contains two copies of an early gene encoding a soluble secreted form of the type II TNF receptor. *Virology*, **204**, 343.

54. Goebel, S. J., Johnson, G. P., Perkus, M. E., Davis, S. W., Winslow, J. P., and Paoletti, E. (1990) The complete DNA sequence of vaccinia virus. *Virology*, **179**, 247.

55. Alcamí, A., Khanna, A., Paul, N. L., and Smith, G. L. (1999) Vaccinia virus strains Lister, USSR and Evans express soluble and cell surface tumour necrosis factor receptors. *J. Gen. Virol.*, **80**, 949.

56. Smith, C. A., Hu, F. Q., Smith, T. D., Richards, C. L., Smolak, P., Goodwin, R. G., *et al.* (1996) Cowpox virus genome encodes a second soluble homologue of cellular TNF receptors, distinct from CrmB, that binds TNF but not LT alpha. *Virology*, **223**, 132.

57. Aguado, B., Selmes, I. P., and Smith, G. L. (1992) Nucleotide sequence of 21.8 kbp of variola major virus strain Harvey and comparison with vaccinia virus. *J. Gen. Virol.*, **73**, 2887.

58. Loparev, V. N., Parsons, J. M., Knight, J. C., Panus, J. F., Ray, C. A., Buller, R. M., *et al.* (1998) A third distinct tumor necrosis factor receptor of orthopoxviruses. *Proc. Natl. Acad. Sci. USA*, **95**, 3786.

59. Tidona, C. A. and Darai, G. (1997) The complete DNA sequence of lymphocystis disease virus. *Virology*, **230**, 207.

60. Smith, G. L. and Chan, Y. S. (1991) Two vaccinia virus proteins structurally related to the interleukin-1 receptor and the immunoglobulin superfamily. *J. Gen. Virol.*, **72**, 511.

61. McMahan, C. J., Slack, J. L., Mosley, B., *et al.* (1991) A novel IL-1 receptor, cloned from B cells by mammalian expression, is expressed in many cell types. *EMBO J.*, **10**, 2821.

62. Alcamí, A. and Smith, G. L. (1992) A soluble receptor for interleukin-1 beta encoded by vaccinia virus: a novel mechanism of virus modulation of the host response to infection. *Cell*, **71**, 153.

63. Spriggs, M., Hruby, D. E., Maliszewski, C. R., Pickup, D. J., Sims, J. E., Buller, R. M. L., *et al.* (1992) Vaccinia and cowpox viruses encode a novel secreted interleukin-1 binding protein. *Cell*, **71**, 145.

64. Symons, J. A., Young, P. R., and Duff, G. W. (1995) Soluble type II interleukin 1 (IL-1) receptor binds and blocks processing of IL-1β precursor and loses affinity for IL-1 receptor antagonist. *Proc. Natl. Acad. Sci. USA*, **92**, 1714.

65. Alcamí, A. and Smith, G. L. (1996) A mechanism for the inhibition of fever by a virus. *Proc. Natl. Acad. Sci. USA*, **93**, 11029.

66. Upton, C., Mosssman, K., and McFadden, G. (1992) Encoding of a homolog of the IFN-γ receptor by myxoma virus. *Science*, **258**, 1369.

67. Alcamí, A. and Smith, G. L. (1995) Vaccinia, cowpox and camelpox viruses encode soluble gamma interferon receptors with novel broad species specificity. *J. Virol.*, **69**, 4633.

68. Blanchard, T. J., Alcamí, A., Andrea, P., and Smith, G. L. (1998) Modified virus Ankara undergoes limited replication in human cells and lacks several immunomodulatory proteins: implications for use as a human vaccine. *J. Gen. Virol.*, **79**, 1159.

69. Mossman, K., Upton, C., Buller, R. M., and McFadden, G. (1995) Species specificity of ectromelia virus and vaccinia virus interferon-gamma binding proteins. *Virology*, **208**, 762.

70. Puehler, F., Weining, K. C., Symons, J. A., Smith, G. L., and Staeheli, P. (1998) Vaccinia virus-encoded cytokine receptor binds and neutralizes chicken interferon-gamma. *Virology*, **248**, 231.

71. Dalton, D. K., Pitts-Meek, S., Keshav, S., Figari, I. S., Bradley, A., and Stewart, T. A. (1993) Multiple defects of immune cell function in mice with disrupted interferon-γ genes. *Science*, **259**, 1739.

72. Huang, S., Hendriks, W., Althage, A., Hemmi, S., Bluethmann, H., Kamijo, R., *et al.* (1993) Immune response in mice that lack the interferon-γ receptor. *Science*, **259**, 1742.

73. Mossman, K., Nation, P., Macen, J., Garbutt, M., Lucas, A., and McFadden, G. (1996) Myxoma virus M-T7, a secreted homolog of the interferon-gamma receptor, is a critical virulence factor for the development of myxomatosis in European rabbits. *Virology*, **215**, 17.

74. Ueda, Y., Morikawa, S., and Matsuura, Y. (1990) Identification and nucleotide sequence of the gene encoding a surface antigen induced by vaccinia virus. *Virology*, **177**, 588.

75. Symons, J. A., Alcamí, A., and Smith, G. L. (1995) Vaccinia virus encodes a soluble type I interferon receptor of novel structure and broad species specificity. *Cell*, **81**, 551.

76. Colamonici, O. R., Domanski, P., Sweitzer, S. M., Larner, A., and Buller, R. M. (1995) Vaccinia virus B18R gene encodes a type I interferon-binding protein that blocks interferon alpha transmembrane signaling. *J. Biol. Chem.*, **270**, 15974.

77. Vancova, I., La Bonnardiere, C., and Kontsek, P. (1998) Vaccinia virus protein B18R inhibits the activity and cellular binding of the novel type interferon-delta. *J. Gen. Virol.*, **79**, 1647.

78. Liptakova, H., Kontsekova, E., Alcami, A., Smith, G. L., and Kontsek, P. (1997) Analysis of an interaction between the soluble vaccinia virus-coded type I interferon (IFN)-receptor and human IFN-alpha1 and IFN-alpha2. *Virology*, **232**, 86.

79. Massung, R. F., Jayarama, V., and Moyer, R. W. (1993) DNA sequence analysis of conserved and unique regions of swinepox virus: identification of genetic elements supporting phenotypic observations including a novel G protein-coupled receptor homologue. *Virology*, **197**, 511.

80. Cao, J. X., Gershon, P. D., and Black, D. N. (1995) Sequence analysis of HindIII Q2 fragment of capripoxvirus reveals a putative gene encoding a G-protein-coupled chemokine receptor homologue. *Virology*, **209**, 207.

81. Lalani, A. S., Graham, K., Mossman, K., Rajarathnam, K., Clark-Lewis, I., Kelvin, D., *et al.* (1997) The purified myxoma virus gamma interferon receptor homolog M-T7 interacts with the heparin-binding domains of chemokines. *J. Virol.*, **71**, 4356.

82. Graham, K. A., Lalani, A. S., Macen, J. L., *et al.* (1997) The T1/35kDa family of poxvirus-secreted proteins bind chemokines and modulate leukocyte influx into virus-infected tissues. *Virology*, **229**, 12.

83. Alcamí, A., Symons, J. A., Collins, P. D., Williams, T. J., and Smith, G. L. (1998) Blockade of chemokine activity by a soluble chemokine binding protein from vaccinia virus. *J. Immunol.*, **160**, 624.

84. Smith, C. A., Smith, T. D., Smolak, P. J., *et al.* (1997) Poxvirus genomes encode a secreted, soluble protein that preferentially inhibits beta chemokine activity yet lacks sequence homology to known chemokine receptors. *Virology*, **236**, 316.

85. Martinez-Pomares, L., Thompson, J. P., and Moyer, R. W. (1995) Mapping and investigation of the role in pathogenesis of the major unique secreted 35-kDa protein of rabbitpox virus. *Virology*, **206**, 591.

86. Strockbine, L. D., Cohen, J. I., Farrah, T., Lyman, S. D., Wagener, F., DuBose, R. F., *et al.* (1998) The Epstein–Barr virus BARF1 gene encodes a novel, soluble colony-stimulating factor-1 receptor. *J. Virol.*, **72**, 4015.

87. Essani, K., Chalasani, S., Eversole, R., Beuving, L., and Birmingham, L. (1994) Multiple anti-cytokine activities secreted from tanapox virus-infected cells. *Microb. Pathol.*, **17**, 347.

88. Neote, K., DiGregorio, D., Mak, J. Y., Horuk, R., and Schall, T. J. (1993) Molecular cloning, functional expression and signaling characteristics of a C-C chemokine receptor. *Cell*, **72**, 415.

89. Gao, J. L., Kuhns, D. B., Tiffany, H. L., McDermott, D., Li, X., Francke, U., *et al.* (1993) Structure and functional expression of the human macrophage inflammatory protein 1 alpha/RANTES receptor. *J. Exp. Med.*, **177**, 1421.

90. Gao, J. L. and Murphy, P. M. (1994) Human cytomegalovirus open reading frame US28 encodes a functional beta chemokine receptor. *J. Biol. Chem.*, **269**, 28539.

91. Kuhn, D. E., Beall, C. J., and Kolattukudy, P. E. (1995) The cytomegalovirus US28 protein binds multiple CC chemokines with high affinity. *Biochem. Biophys. Res. Commun.*, **211**, 325.

92. Michelson, S., Dal Monte, P., Zipeto, D., Bodaghi, B., Laurent, L., Oberlin, E., *et al.* (1997) Modulation of RANTES production by human cytomegalovirus infection of fibroblasts. *J. Virol.*, **71**, 6495.

93. Bodaghi, B., Jones, T. R., Zipeto, D., Vita, C., Sun, L., Laurent, L., *et al.* (1998) Chemokine sequestration by viral chemoreceptors as a novel viral escape strategy: withdrawal of chemokines from the environment of cytomegalovirus-infected cells. *J. Exp. Med.*, **188**, 855.

94. Vieira, J., Schall, T. J., Corey, L., and Geballe, A. P. (1998) Functional analysis of the human cytomegalovirus US28 gene by insertion mutagenesis with the green fluorescent protein gene. *J. Virol.*, **72**, 8158.

95. Davis-Poynter, N. J., Lynch, D. M., Vally, H., Shellam, G. R., Rawlinson, W. D., Barrell, B. G., *et al.* (1997) Identification and characterization of a G protein-coupled receptor homolog encoded by murine cytomegalovirus. *J. Virol.*, **71**, 1521.

96. Levy, J. A. (1997) Three new human herpesviruses (HHV6, 7 and 8). *Lancet*, **349**, 558.

97. Isegawa, Y., Ping, Z., Nakano, K., Sugimoto, N., and Yamanishi, K. (1998) Human herpesvirus 6 open reading frame U12 encodes a functional beta-chemokine receptor. *J. Virol.*, **72**, 6104.

98. Bais, C., Santomasso, B., Coso, O., *et al.* (1998) G-protein-coupled receptor of Kaposi's sarcoma-associated herpesvirus is a viral oncogene and angiogenesis activator. *Nature*, **391**, 86.

99. Geras-Raaka, E., Varma, A., Ho, H., Clark-Lewis, I., and Gershengorn, M. C. (1998) Human interferon-gamma-inducible protein 10 (IP-10) inhibits constitutive signaling of Kaposi's sarcoma-associated herpesvirus G protein-coupled receptor. *J. Exp. Med.*, **188**, 405.

100. Ahuja, S. K. and Murphy, P. M. (1993) Molecular piracy of mammalian interleukin-8 receptor type B by herpesvirus Saimiri. *J. Biol. Chem.*, **268**, 20691.

101. Preas, H. L., Reda, D., Tropea, M., Vandivier, R. W., Banks, S. M., Agosti, J. M., *et al.* (1996) Effects of recombinant soluble type I interleukin-1 receptor on human inflammatory responses to endotoxin. *Blood*, **88**, 2465.

102. Margulies, B. J., Browne, H., and Gibson, W. (1996) Identification of the human cytomegalovirus G-protein coupled receptor homologue encoded by UL33 in infected cells and envelope virus particles. *Virology*, **225**, 111.

103. Virgin, H. W., Latreille, P., Wamsley, P., Hallsworth, K., Weck, K. E., Dal Canto, A. J., *et al.* (1997) Complete sequence and genomic analysis of murine gammaherpesvirus 68. *J. Virol.*, **71**, 5894.

104. Telford, E. A., Watson, M. S., Aird, H. C., Perry, J., and Davidson, A. J. (1995) The DNA sequence of equine herpesvirus 2. *J. Mol. Biol.*, **249**, 520.

7 | Genetic variation in cytokines and relevance to inflammation and disease

GORDON DUFF

1. Introduction

Gene mutations that give a reproductive advantage will be propagated in a population. Mutations in genes involved in defence against infection will be selected if they increase the chances of their owners surviving to reproductive age.

Cytokines provide intercellular signals to co-ordinate and control multicellular co-operation and recent work has shown a high degree of polymorphism in the cytokine genes involved in inflammation and immunity. Polymorphisms are defined as mutations in DNA that have reached a frequency of 2% in a population. Cytokine gene polymorphisms (DNA sequence variations) are frequently in regions of the gene that regulate transcription or post-transcriptional events, and so can be functionally significant. The inflammatory response, therefore, is genetically programmed both quantitatively and qualitatively, with some people having a very vigorous response and others a more measured response to the same stimulus.

The inflammatory response is primarily homeostatic, achieving a fine balance that eradicates pathogens and repairs injured tissues while limiting the cost to the host. But the regulation can fail in circumstances of which we have, in most cases, a very limited understanding. When this happens, recurrent and prolonged inflammatory responses occur, often in the absence of a detectable triggering infection. If this continues, tissues and organs become irreversibly damaged, leading to chronic inflammatory diseases that have a great impact on public health.

Common pathological features of these cytokine-related conditions include inflammatory cell infiltration of target organs, loss of normal cellular components, and tissue-matrix damage. Such diseases are often chronic, being more or less lifelong once established, and they have a tendency to occur in families. The genetic basis for the familial tendencies of common inflammatory diseases has not been defined fully in any particular case. However, it seems likely, from epidemiological studies and evidence from monozygotic and dizygotic twins, that genetic factors contribute to disease susceptibility and the clinical severity, although environmental factors are also important or essential for the development of clinical disease. Thus, common familial diseases appear to be multifactorial with the involvement of an unknown number of genes and unknown environmental factors.

In many cases, associations between inflammatory, infectious, or immune diseases and alleles, or extended haplotypes, of the major histocompatibility complex (MHC) have been established. These associations have mostly been interpreted as 'immune response' gene effects and, more recently, in terms of the genetically defined ability of MHC molecules to accommodate peptides for antigen presentation to T cells. This mechanism could explain the immunopathogenic component of many familial diseases.

Within the MHC, diseases are often associated with quite extensive haplotypes, making it difficult to distinguish which particular allele(s) may contribute to patho-genesis and which may be associated with disease through physical linkage with the true aetiological gene (linkage disequilibrium). Not only can it be difficult to assess the contribution of an individual allele within an MHC haplotype, but it seems certain that susceptibility to, or severity of, many diseases is also influenced by genes located outside the MHC. Progress in defining these non-MHC genes has been, until recently, somewhat slow.

Recent work has tied several important protozoal and bacterial diseases and many idiopathic chronic inflammatory diseases to the genetic programming of the inflam-matory response (1). Genetic predictors of a large response also predict a higher risk of developing common inflammatory diseases such as rheumatoid arthritis, asthma, psoriasis, ulcerative colitis, periodontitis, and complications of diabetes, as well as other diseases with less obvious inflammatory mechanisms such as coronary heart disease and post-menopausal osteoporosis. These findings may contribute to a mech-anistic explanation of some surprising epidemiological observations at a population level, such as the association between periodontitis and coronary heart disease.

In this chapter, genetic polymorphisms in some of the major inflammatory response cytokines will be reviewed. Research in recent years has produced a very large literature on this topic and since it is impossible to cover all aspects in a chapter of this scope, a large degree of selection has been necessary.

2. Cytokines as candidate genes

There are several approaches to the identification of disease-associated genes (2). When the mechanism of disease is unknown and there are no clues to probable con-tributory genes, a genome-wide approach is essential. Linkage analysis of the disease phenotype within families may identify first the chromosome and then the chromo-somal region where disease genes are located. The human genome map now com-prises so many well-spaced polymorphic markers that linkage analysis can relatively easily locate disease genes in monogenic diseases with a high degree of success. Clearly, linkage analysis is less sensitive in the analysis of multigenic and multi-factorial diseases unless major single-gene effects are present. However, several successful studies have been performed where the clinical resource has been large enough to provide sufficient statistical power.

Another approach is to propose putative genes that may be involved in pathogenesis, termed 'candidate genes'. This is often possible when there is some understanding of the disease process or when linkage studies have already identified the chromosomal region containing contributory genes. Thus, a 'candidate' gene approach is based on forming a hypothesis that can be tested. The macrophage-derived cytokines that control the inflammatory response would seem to be reasonable 'candidate genes' in inflammatory diseases. Not only are they active in the pathogenesis of many such diseases (3, 4) but they also show stable interindividual differences in rates of production (5).

To test the hypothesis that a specified gene may contribute to a disease, linkage analysis in families, disease association studies in populations of unrelated individuals, and family-based association studies, such as the transmission disequilibrium test (TDT), can all be used depending on the clinical resource available.

Association studies can be very sensitive and single nucleotide polymorphisms (SNPs) are usually optimal when the heterozygosity rate is high enough. A single base change gives rise to three genotypes: homozygosity for the commoner form or allele, usually described as 1/1; homozygosity for the rarer allele, 2/2; and heterozygosity with a different allele on each chromosome, 1/2. Conventionally, the commonest allele of a polymorphism is called 1, the next commonest 2, and so on. An asterisk is often used to denote an allele, e.g. *1 would be the commonest allele of the system in question.

Although sensitive, association studies in populations of unrelated individuals are liable to give spurious results. Genetic admixture in the population is a major source of spurious association; the population size may be too small to provide adequate statistical power; there may be a high degree of genetic heterogeneity in a disease phenotype; or there may be variability of clinical criteria in a multicentre study, and the practice of making multiple comparisons in the same population can give rise to spuriously significant P-values unless a correction is made. For these reasons, association studies require large populations (large clinical resource), standardized clinical phenotyping, avoidance of ethnic or racial heterogeneity in the population, and replication in independent populations. Newer methods using family members to provide the control population, such as the TDT, can overcome at least some of these problems (2).

Whichever method is used, it is necessary to identify polymorphic markers within or around the gene of interest. In an association study the allelic frequency (gene frequency) is compared between a population of unrelated individuals with the disease and a relevant healthy control population. If a disease-associated allele is identified in such a population analysis, a further question is raised: does the identified gene, itself, contribute to the disease process or is it physically linked on the chromosome with the contributory gene, so acting as a 'marker' for it? If a candidate gene has been proposed on the basis of a biological role in the disease and is then found to possess a disease-associated allele, its likelihood of being a contributory gene may be greater. This would require that the polymorphism resulted in an altered protein, or in altered regulation of the gene. The latter might result in a quantitative difference in gene regulation (e.g. increased rate of gene transcription) or a qualitative difference in gene regulation (e.g. a change in gene transcription in response to a specific signalling pathway or in a particular cell type).

Eukaryotic RNA contains sequences that can be spliced together as mRNA (exons) most of which are usually translated into protein. There are non-coding sequences (introns) between the exons. The DNA 'upstream' of the first exon (5' flanking region) contains short nucleotide sequences that bind transcription factors and thereby control the activity of RNA polymerase and the process of gene transcription (the promoter region). After the splicing out of introns, not all of the mRNA product is translated into protein. Whether it is translated or not, it can affect the production of protein; for example, untranslated sequences at the 3' flanking region influence the stability of the mRNA of many cytokine genes. Short DNA sequences in the 5' or 3' flanking regions or anywhere within the gene itself, or at some distance, can act to increase gene transcription (enhancer sequences) or to reduce it (repressor sequences).

3. TNF locus polymorphisms and the MHC

The search for disease-related polymorphisms in cytokine genes is relatively recent. Because of its biological properties and its location within the MHC, there was early interest in potential disease-related alleles of the tumour necrosis factor (TNF) locus, and the first such observation was made in mice (6). A restriction fragment length polymorphism (RFLP) within the TNF-α gene of lupus-prone (NZB $\times$ NZW) F_1 mice correlated with reduced production of TNF-α and was thought to be related to the development of the lupus-like nephritis in these mice (6).

In humans, early attempts to find polymorphisms in the TNF region that might be associated with MHC haplotypes met with mixed success (7–9). It was possible to relate an *Nco*I RFLP at the TNF locus to HLA-B and a DR haplotype (7). This RFLP was also related to modest changes in TNF production rates *in vitro* (10). Associations between HLA types and TNF-α production rate *in vitro* have also been noted (11, 12). Interesting potential disease associations with these markers, for example with multiple sclerosis and optic neuritis (13), were reported. In fact, the *Nco*I RFLP at the TNF locus was later found to be located within the first intron of the TNF-β gene (lymphotoxin) (14, 15). This polymorphism was shown to be correlated with an amino acid variation at position 26 of TNF-β and with a reduced rate of TNF-β production (15).

The search for further and possibly more informative polymorphisms at the TNF locus continued with the identification of four dinucleotide repeats of variable length (microsatellites). Of these, two were 3.5 kb upstream of the TNF-β gene, one was some 10 kb downstream of the TNF-α gene, and one TC repeat was located within the first intron of the TNF-β gene (16). Although the previously discussed polymorphisms are within the TNF-β gene, or at some distance from the tandem TNF genes, polymorphisms were also found within the promoter region of the murine TNF-α gene (17) and at position −308 (18) and −238 (19) within the promoter region of the human TNF-α gene. Polymorphism across the 3′ untranslated region of TNF-α, in translational repressor sequences, has also been described in mice (20), but not in a human population (21).

3.1 TNF locus polymorphism and diseases

Several early studies tested whether there were disease-associated or functionally different alleles of the TNF locus. No association was found between a lymphotoxin polymorphism (TNF-β) and the low TNF-β production rates seen in patients with primary biliary cirrhosis (22) and, likewise, there was no relation between lymphotoxin polymorphism and ankylosing spondylitis (23). Associations were found between TNF microsatellites and other loci within the MHC including Class I, II, and III alleles (24).

A single nucleotide polymorphism (SNP) at −308 within the human TNF-α promoter was found to be highly associated with the 'autoimmune haplotype' HLA-A1, B8, DR3, DQ2 (25). The lymphotoxin gene *NcoI* RFLP also correlated with MHC ancestral haplotypes including HLA-A1, B8, and DR3 (26, 27). This HLA haplotype has also been associated with increased TNF-α production by lymphoid cell lines *in vitro* (27, 28). Because the TNF −308 polymorphism is in linkage disequilibrium with other MHC genes of the region, producing common haplotypes, and with the TNF-β polymorphism, it is not surprising that TNF alleles have been found at raised frequencies in diseases associated with the HLA-A1, B8, DR3, DQ2 haplotype. For example, the TNF-β RFLP has been associated with Graves' disease of the thyroid (29) and also with systemic lupus erythematosis (SLE) (30). An association has also been found between SLE and the TNF-α −308 base transition (31). With this polymorphism, the association was even stronger between TNF-α rare allele and the presence of anti-Ro and anti-La autoantibodies (31).

In these studies, however, the disease association was usually stronger with HLA DR3 than with TNF, suggesting that the TNF association resulted from linkage disequilibrium between the TNF locus and the DR3 locus or DQ2 locus (32). A possible exception is in coeliac disease (33) where the association was stronger with TNF-α than with any other of the tested HLA alleles. Other such stronger associations are described below. In lupus-prone strains of mice, it was possible to show that disease susceptibility, TNF-α production, and TNF-α genotype were all associated, suggesting a direct involvement of the TNF gene in disease pathogenesis (34). These mice have a promoter region polymorphism within the TNF-α gene.

In reporter gene assays of the human TNF-α promoter performed in lymphocytic cell lines, the rarer allele at −308 was several-fold more efficient than the common promoter region variant (35–37), although one laboratory found no difference between the alleles, perhaps because of different sequences used in the experiments (38). In support of the differential transcriptional findings, TNF-α production in endotoxin-stimulated whole blood cultures was reported to be increased in individuals who carry –308*2, the higher transcription allele (39), and stimulation of mononuclear cells with anti-CD3 antibody gave a similar result (40).

There have been several reports of associations between infectious diseases and the –308 SNP of TNF-α. The most striking association with the TNF-α gene was the finding, in the Gambia, that children who were homozygous for –308*2 were eightfold more likely to have a severe clinical outcome of cerebral malaria (41). Also in the Gambia, a strong association was found between scarring trachoma, TNF levels in tear fluid, and –308*2 (42). The genetic association with TNF appeared to be independent of other HLA alleles. The same TNF allele, TNF –308*2, appeared to predispose to leishmaniasis (43), worse clinical outcomes in meningococcal infection (44), to lepromatous leprosy (45), and a linked marker predisposed to post-traumatic sepsis (46). Conversely, –308*2 was apparently protective in chronic human cytomegalovirus infection (47), perhaps indicating that a more vigorous inflammatory response against this virus is beneficial to the host.

In other non-infectious diseases, the TNF –308 polymorphism was significantly associated with the following: pre-eclampsia (48); SLE in African-Americans, independently of DR alleles (49); insulin resistance and raised serum leptin levels (50); obesity (–308*1) (51); chronic lymphatic leukaemia (52); non-Hodgkin lymphoma (53); and treatment outcome in non-Hodgkin lymphoma (54). There have been several reports of –308 association with respiratory diseases: asthma (55); childhood asthma (–308*1) (56); chronic bronchitis (57); and acute sarcoidosis (Lofgren syndrome) (58). In liver diseases, the –308 polymorphism has been implicated in primary biliary cirrhosis (59) and in primary sclerosing cholangitis (60). In myasthenia gravis, reports suggest an association between TNF locus polymorphism and thymoma (61) and with young age of onset (62). The –308*2 frequency was significantly raised in patients with more frequent heart transplant rejection episodes (63) and in quartz dust-related scleroderma (64). No overall association was found in rheumatoid arthritis (65) but an association with the nodular subgroup of this disease has been reported (66).

Disease associations have also been reported with the –238 SNP of TNF-α. The same disease associations often occur for –308 and –238, but although they are physically very close together, associations with –238 on occasion seem independent of the –308 SNP and vice versa. There is, as yet, no evidence for any functional significance of the TNF SNP –238 (19, 67). Reported associations with –238 include juvenile-onset psoriasis and psoriatic arthritis (68, 69); joint damage in rheumatoid arthritis (67); and chronic hepatitis B infection (70). Associations with both –308 and –238 have been reported in HLA B27-positive patients with ankylosing spondylitis (rare alleles being protective) (71).

Although there are many outstanding issues to be resolved, it now seems likely that:

(a) Polymorphisms at the TNF locus form part of extended MHC haplotypes such as the A1, B8, DR3, DQ2 haplotype that has been associated with a range of autoimmune human diseases.

(b) TNF polymorphisms may contribute to disease susceptibility independently of linked genes in the MHC.

(c) The genetic significance of TNF may be especially important in infectious diseases, but may also apply to a wide range of apparently non-infectious diseases where the balance of the inflammatory response affects the clinical outcome.

(d) The promoter SNP at −308 is associated with functional changes in TNF gene regulation, which, in turn, alter TNF protein production in different cells and possibly in response to different signals.

4. The interleukin 1 gene cluster

The genes of the interleukin-1 (IL-1) cluster, comprising IL-1α, IL-1β, and IL-1 receptor antagonist (IL-1ra) have all been mapped to a 430 kb stretch of DNA on the long arm of human chromosome 2 (72). The IL-1 receptor Type I (and possibly IL-1 receptor Type II) also map to 2q. The gene products of these loci have been implicated in many infectious and inflammatory diseases.

Several different polymorphisms have been found and characterized within the IL-1 gene cluster. They include single base changes in 5′ flanking DNA of IL-1β (73) and IL-1α (74), as well as polymorphic variable number tandem repeats (VNTR polymorphism) in IL-1α (75) and IL-1ra (76, 77). In IL-1α, the VNTR is made up of a 46 bp stretch of DNA that is repeated 5–19 times within intron 6. There are six alleles, of which the most common (62%) has nine repeats and the second most common (23%) has 18 repeats (75). Each repeat contains sites of potential significance in gene regulation, in particular a consensus glucocorticoid response element and a potential binding site for the transcription factor SP-1, and the repeat seems to have repressor function on homologous and heterologous promoters, increasing with the number of repeats present (78). There is also a *Taq*I RFLP in the IL-1 gene cluster that is associated with higher production of IL-β (79).

The IL-1ra is a very powerful anti-inflammatory agent *in vivo* (80). Within the IL-1ra gene there is a VNTR polymorphism in intron 2. This is comprised of an 86 bp tandem repeat and there are five alleles, with four repeats being the commonest allele (74%) and two repeats the next most frequent allele (22%) (77). These repeat stretches, like those in IL-1α, also contain potential binding sites for transcription factors but no molecular function has yet been demonstrated for any of these IL-1 gene cluster polymorphisms. However, current work in several laboratories is beginning to uncover functional differences at the cellular and tissue levels in relation to IL-1 cluster genotypes. For example, in ulcerative colitis, bowel mucosa contains lower amounts of IL-1ra in individuals who are homozygous for allele 2 of the IL-1rn gene (81) and blood levels of IL-1ra in healthy donors were reported to be about twice as high in carriers of the IL-1rn*2 allele (82). However, using an allele-specific measurement of mRNA in heterozygous cells (based on exonic SNPs in linkage with the VNTR), no significant differences were found (83). A high degree of linkage disequilibrium exists across the IL-1 gene region, roughly in relation to distance, and it has been possible to define IL-1 cluster haplotypes based on eight markers (84). The functional significance, if any, of particular polymorphisms should, therefore, be considered in relation to the alleles of the other IL-1 genes with which it is in linkage disequilibrium. This is particularly important in the IL-1 cluster since the protein products of the three genes act as both agonists and antagonists of the same cellular receptor.

4.1   IL-1 locus polymorphisms and diseases

Several associations have been found between polymorphisms of the IL-1 gene cluster and a range of inflammatory diseases: lichen sclerosus (85), alopecia areata (86), SLE (87), diabetic kidney disease (nephropathy) (88), ulcerative colitis (89–92), juvenile chronic arthritis (93), and psoriasis (94). Not all of these findings have been replicated in other studies; for example, the association between IL-1rn*2 and ulcerative colitis (95). Different populations, particularly with respect to ethnicity, may contribute to this, as may underpowered studies on relatively small cohorts, but there may also be real genetic differences in different populations.

The association between IL-1rn and diabetic nephropathy (88) has been supported by reports of association of the same allele with other renal diseases such as nephropathy in Henoch–Schonlein purpura (96) and recurrent haematuria in IgA nephropathy (97), and a report of an association between IL-1β and diabetic nephropathy (98). An association between Type I diabetes itself and both the IL-1rn and the IL-1β genes has also been observed (99). Recently a composite genotype involving IL-1β +3953 was found to be significantly over-represented in patients with severe periodontal disease (odds ratio = 18 for 40–60 age group) (100). This is of interest because IL-1β has been strongly implicated as a mediator of tissue resorption in periodontitis and the +3953 allele that is associated with periodontitis severity is the mutation that produces the *Taq*I RFLP, the variant associated with higher levels of IL-1β production *in vitro*. The observation in periodontal disease was replicated in an independent study where it was also noted that the disease-associated genotype also correlated with higher IL-β production from stimulated blood cells of patients (101).

In neurological diseases, one study found a raised frequency of IL-1rn*2 in the relapsing-remitting form of multiple sclerosis (102) while another group observed that IL-1rn*2 and IL-1β*2 seemed to be markers for a faster rate of progression on the disability scale (103). Homozygosity for IL-1β +3953 *2/2 was significantly commoner in myasthenia gravis in one study (odds ratio = 4.6) (104). Apart from the earlier reports of IL-1 associations with the chronic iridocyclitis subset of juvenile chronic arthritis and also with SLE, IL-1rn has also been found to be associated with osteoporosis (105) and with more severe cases of Sjogren's syndrome (106). In the field of infectious diseases, large studies are currently underway, but there has been one interesting observation in Epstein–Barr virus (EBV) seropositivity in a cohort of 400 Finnish blood donors where IL-1rn*2 and IL-1β*2 (i.e. a 'pro-inflammatory' genotype) seemed to correlate with EBV seropositivity (107). Chronic arterial disease has increasingly been recognized as having an inflammatory component in the arterial wall and IL-1 has been found both in the arteries and the myocardium of diseased hearts (108). In a recent large study of angiographically-measured coronary artery occlusion involving two centres, the IL-1rn*2 gave an odds ratio of 2.8 for single vessel disease (109), other large studies in cardiovascular disease are underway.

These associations represent statistically significant differences in allele frequencies between disease populations and matched control populations. The extent to which the observations will be replicated and their generalizability across ethnic groups remains to be seen. What the significance may be in terms of disease pathogenesis is not yet known, but studies have related cytokine production rates to the genetic markers of the IL-1 gene cluster and direct testing *in vitro* of the functional significance of these polymorphisms is currently underway. It is noteworthy that where differences in cytokine levels correlate with a disease-associated genotype, it is often in the relevant inflamed tissue. In many of the clinical diseases where associations with the IL-1 gene cluster have been found, there are usually also many reports that implicate the IL-1 system in the pathogenesis of the disease (1, 4). The Type I IL-1 receptor is also polymorphic but, to date, no studies on its association with disease have been published.

4.2   Susceptibility or severity?

The disease associations that have been established with IL-1 gene cluster polymorphisms have been detected by comparing allele frequencies in disease and in normal populations. In this way it is possible to generate hypotheses about susceptibility genes, which may be normal variants present in a large proportion of the population. This is in contrast to a mutation in a single gene leading to a disease. For example, mutations in the gene that encodes IL-2 receptor α chain 'cause' X-linked severe combined immunodeficiency disease (SCID) (110). Whether genetic linkage analysis will show that other diseases map to single genes encoding cytokines or their receptors also remains for the future.

The data collected to date seem to indicate that IL-1 gene polymorphisms represent markers of disease severity. For example, in ulcerative colitis the overall association with IL-1rn*2 could be accounted for by a stronger association with the subgroup of total colitis, i.e. the most severe clinical phenotype (89). Similarly in periodontal disease the IL-1 composite genotype was most highly over-represented in the patients with severe disease (100). In some ways, the distinction between susceptibility and severity of disease may be semantic, arising from the descriptive nature of disease classification. In other words, mild ulcerative colitis may be a different disease from severe ulcerative colitis in terms of the genetic pathways and environmental triggers that initiate and perpetuate it.

5.   IL-4 and IL-4R polymorphisms and diseases

There have been several studies of polymorphisms in the IL-4 gene and the α chain of the IL-4 receptor in immunological diseases. Because IL-4 signals isotype switching to IgE in B cells, it was considered a candidate gene in atopy. A weak association with some specific IgEs was found with a –590 polymorphism of IL-4, but no association with asthma or with total IgE levels (111). However, in a TDT study of Japanese children, there was evidence of association between –590 and asthma but not with IgE (112). In a recent linkage study in asthma, serum levels of allergen-specific IgEs appeared to be linked with the IL-4 locus but not with the –590 polymorphism (113). An IL-4 VNTR polymorphism has been associated with the age of onset of multiple sclerosis in Italian patients (114), but not with another immune disease, myasthenia gravis (115).

A strong association was found between atopy and an SNP at position 1902 that changes an amino acid (amino acid 576) in the cytosolic domain of the IL-4R α chain (odds ratio = 9) (116). In another recent study, this polymorphism and another at amino acid 503 were related to total serum IgE levels and affected the signalling function of the receptor, as shown by altered phosphorylation of IRS-1 and IRS-2 (117). On the other hand, a study of Japanese patients showed no linkage or association between asthma or atopy and the IL-4R α locus, including the amino acid 576 polymorphism. This underscores the differences between genetic factors in the same diseases in different ethnic and racial groups.

6. IL-6 gene polymorphism and diseases

Restriction fragment length polymorphisms in human (9, 10) and mouse (118) IL-6 genes were described very early in the development of this field but case control studies in disease have been slower to appear. An *rflp* allele arising from variation in the 3′ flanking region of human IL-6 was found at raised frequency in SLE, in patients who all carried the DQβ6 gene in the HLA class II region. In this study, carriage of the same IL-6 allele correlated in affected and unaffected members of a family, with higher constitutive levels of IL-6 mRNA in the cells of family members (119). Later, alleles of a 3′ microsatellite were reported to be associated with bone mass at the lumber spine in both pre- and post-menopausal women (120). The results suggested that the IL-6 contribution might be to peak bone mass, an important variable in the development of clinical osteoporosis.

Recently, a very comprehensive study was made of the IL-6 gene in systemic juvenile chronic arthritis. A promoter region, SNP, was described that altered transcription in reporter gene assays. The poor promoter had a C in place of a G at –174. In patients with the C/C genotype, plasma levels of IL-6 were lower and this genotype was under-represented in the disease population, especially in children below the age of five. Thus, the C/C genotype at –174 of the IL-6 gene conferred poor promoter function, was associated with lower plasma IL-6 levels, and seemed to be protective for JCA, especially in younger children (121). This report is an excellent example of bringing together several different strands of evidence to evaluate the potential significance of a single nucleotide variation in a cytokine promoter.

7. IL-10 gene polymorphism and diseases

Several SNP and microsatellite polymorphisms have been described in the IL-10 gene and its promoter (122–125). Associations have been found with SLE (123, 126). There was evidence for a multiplicative effect in SLE of an IL-10 polymorphism with a polymorphism in the *bcl-2* gene (126). Others have reported associations of IL-10 alleles with heart transplant rejection (combined with TNF-α −308*2) (63); graft-versus-host disease (127); total serum IgE in allergic disorders (125), but there was no association detected in either rheumatoid arthritis or Felty's syndrome (128).

8. Chemokine receptor gene polymorphisms

Much of the work of the last five years on chemokine receptors has been driven, not by interest in cytokine signalling, but by the discovery that CC and CXC chemokine receptors act as co-receptors on CD4$^+$ cells for HIV-1. A large literature now exists and it is beyond the scope of this chapter to summarize it comprehensively. However, the main points relevant to the topic of cytokine polymorphism are described below.

It had been known from the mid 1990s that chemokines could retard the *in vitro* infectivity of T cells by HIV and evidence emerged that chemokine receptors such as CCR5 could act as co-receptors for HIV (129). Chemokines such as RANTES and MIP-1α blocked the cellular entry of HIV independently of the signalling function of receptors such as CCR5 (130, 131). Furthermore, the V3 domain of HIV-1 gp120 envelope glycoprotein was involved in chemokine inhibition of infectivity (132).

In 1996, the observation was made that cells derived from two individuals who had had multiple exposure to HIV, but remained uninfected, were homozygous for a 32 bp deletion mutation of the chemokine receptor CCR5 (132). The deletion was in the structural gene and rendered CCR5 undetectable on cell surfaces. Then, in a large cohort of 1252 homosexual males in San Francisco, it was noted that homozygosity for the CCR5 32 bp deletion had a highly protective effect against HIV infection, with some protection against disease progression being seen in heterozygotes (133).

Since then many reports have appeared and it seems clear that the CCR5 deletion is not the only factor in HIV resistance related to chemokines (134). Other chemokine receptors, such as CXCR4 and CCR2, can act as co-receptors for HIV, probably depending on the HIV subtype. In a study of long-term, HIV-positive non-progressors, heterozygosity for the CCR5 deletion was shown to be neither sufficient nor essential for protection against AIDS progression (135) but homozygosity conferred resistance to infection. In a study of males and females, heterozygosity for the CCR5 deletion polymorphism seemed to be protective for heterosexual but not homosexual partners of HIV-infected individuals (136). It has been shown that gp120 binding to CD4 recruits CXCR4 into a tri-molecular complex prior to cellular entry of virus (137), confirming the role of chemokine receptors, in addition to CCR5, as potential co-receptors for the HIV virus.

A promoter, SNP, in the CCR5 gene was associated with altered transcriptional function, the G allele was a weaker promoter in reporter assays and also, in a population study, was protective for HIV disease (138). Conversely, homozygosity for a multisite haplotype of the CCR5 gene containing a promoter allele, CCR5P1, was associated with more rapid progression to AIDS following HIV infection (139). It was estimated that 7–13% of the population have this homozygous genotype. Another interesting polymorphism was found in the 3′ region of the stromal-derived factor gene (SDF-1), this is a ligand for CXCR4, and homozygosity for the A allele of the SNP showed a strong protective effect against AIDS in a large, predominantly US, population of 2857 patients (140). Thus the HIV interaction with chemokine receptors has provided a clear message on the host genetic component in HIV infection and AIDS progression, and the importance of cytokine or cytokine receptor polymorphisms in susceptibility to infectious diseases. The CCR5 deletion is beginning to be studied in other chronic inflammatory diseases and associations are being found. For example, in rheumatoid arthritis, the deletion allele was at a raised frequency in the subgroup lacking IgM rheumatoid factors and also seemed to correlate with a less severe clinical course (141). We can expect to hear more in the future of chemokine receptor polymorphism in other diseases apart from AIDS.

9. Conclusion

The discovery in recent years of polymorphisms in cytokine and related genes has made it possible to test allelic associations with many common diseases. In the case of the pro-inflammatory cytokine TNF-α, the picture is complicated because of haplotypic association with other highly polymorphic and abundant genes of the MHC. An allele of TNF-α is associated with the MHC haplotype A1, B8, DR3, DQ2, which is itself associated with several autoimmune diseases such as SLE, Type I diabetes, and Graves' disease. TNF-α, therefore, is also associated genetically with these diseases but it is more difficult to isolate the effect of the single gene from that of the extended haplotype. However, there is accumulating evidence that TNF-α, itself, is a contributory gene in some diseases, particularly infectious diseases.

The IL-1 gene cluster on chromosome 2 also has several polymorphic markers and it has been possible to find disease-associated alleles in many chronic inflammatory diseases such as SLE, ulcerative colitis, psoriasis and juvenile chronic arthritis, periodontal disease, arterial disease, and even in osteoporosis. IL-1α, IL-1β, and IL-1ra have all been implicated in the pathogenesis of these disorders but it is still possible that the polymorphisms in this gene cluster are markers for disease-related genes in linkage disequilibrium, and other genes within this cluster are being discovered. Studies are in progress to resolve this issue and also to test directly the effect of polymorphisms on gene function and the production rate of cytokine in different cell types and disease states.

For many years, the genes of the MHC were the focus of attention for investigators interested in the genetic basis of immune-related diseases. It would now appear that genetic analysis of the cytokine system will contribute to the understanding of these conditions and also of the host response to cancer and infection, such as TNF-α in malaria and CCR5 in AIDS.

In the past, 'genetic diseases' were those in which a mutation in a critical gene led directly to disease pathogenesis. With the progress in molecular genetics of the last decade or so, we can now extend the idea of 'genetic diseases' to those common conditions in which DNA variation in the normal human population contributes to susceptibility or to the severity of disease and its clinical outcome. The clinical potential to select specific therapeutic approaches seems clear.

The relative risk of disease conferred by cytokine genotype is often high enough to be of immediate clinical utility in genetic diagnostics. For example, in the perio-dontitis studies, an IL-1 genotype conferred a relative risk of 8 for severe disease, and an IL-1rn genotype carried a relative risk of 2.8 for single vessel coronary artery disease (about the same as raised cholesterol). This information can be used, in combination with other known risk factors, to guide treatment decisions and, in the context of preventive medicine, it could ultimately be used to devise risk-reduction regimes for healthy 'at risk' individuals.

It is appropriate to end this chapter on a recent note from genetic anthropology. The CCR5 32 deletion genotype was ascertained in 4166 individuals across the world in an international study (142). A cline of allele frequencies from 0–14% was found across Eurasia. The variant was absent in African, American-Indian, and East Asian populations. Haplotype analysis allowed an estimate to be made of the origin of this mutation to around 700 years ago in north Europe. The cline of gene frequencies and the recent emergence suggest a strong selective event such as an epidemic caused by a microbe that, like HIV, utilizes CCR5 on human cells. Bubonic plague could be a contender for this role.

It seems that gene mutations in our forebears, by adapting the inflammatory response, had survival value against the lethal infections of the time. However, microbes mutate too and a host mutation conferring resistance to a specific infection may leave the door open for another (or future) pathogen. In today's relatively sanitized world, we pay a price for resistance to infections in past generations in terms of susceptibility to diseases arising from uncontrolled inflammation in the absence of any identifiable infection. The genetic perspective may open the way to a better understanding of inflammation and offers new approaches to risk-assessment, treatment, and prevention of inflammatory diseases.

References

1. Cox, A. and Duff, G. W. (1996) Cytokines as genetic modifying factors in immune and inflammatory diseases. *J. Ped. Endocrinol. Metab.*, **9**, 129.
2. Risch, N. and Merikangas, K. (1996) The future of genetic-studies of complex human-diseases. *Science*, **273**, 1516.
3. Duff, G. W. (1989) Peptide regulatory factors in non-malignant disease. *Lancet*, **i**, 1432.
4. Dinarello, C. A. and Wolff, S. M. (1993) The role of interleukin-1 in disease. *N. Engl. J. Med.*, **328**, 106.

5. Molvig, J., Back, L., Cristensen, P., *et al.* (1998) Endotoxin-stimulated human monocyte secretion of interleukin-1 tumor necrosis factor alpha, and prostaglandin E2 shows stable interindividual differences. *Scand. J. Immunol.*, **27**, 705.

6. Jacob, C. O. and McDevitt, H. O. (1988) Tumor necrosis factor-alpha in murine auto-immune 'lupus' nephritis. *Nature*, **331**, 356.

7. Choo, S. Y., Speis. T., Strominger, J. L., and Hansen, J. (1988) Polymorphism in the tumor necrosis factor gene: association with HLA-B and DR haplotypes. *Hum. Immunol.*, **23**, 86.

8. Partanen, J. and Koskimies, S. (1988) Low degree of DNA polymorphism in the HLA-linked lymphotoxin (tumour necrosis factor beta) gene. *Scand. J. Immunol.*, **28**, 313.

9. Fugger, L., Morling, N., Ryder, L. P., *et al.* (1989) Ncol restriction fragment length polymorphism (RFLP) of the tumor necrosis factor (TNF alpha) region in primary biliary cirrhosis and in healthy Danes. *Scand. J. Immunol.*, **30**, 185.

10. Fugger, L., Bendtzen, K., Morling, N., *et al.* (1989) Possible correlation of TNF alpha-production with RFLP in humans [letter]. *Eur. J. Haematol.*, **43**, 255.

11. Bendtzen, K., Morling, N., Fomsgaard, A., *et al.* (1988) Association between HLA-DR2 and production of tumour necrosis factor alpha and interleukin-1 by mononuclear cells activated by lipopolysaccharide. *Scand. J. Immunol.*, **28**, 599.

12. Jacob, C. O., Lewis, G. D., and McDevitt, H. O. (1991) MHC class II-associated variation in the production of tumor necrosis factor in mice and humans: relevance to the pathogenesis of autoimmune diseases. *Immunol. Res.*, **10**, 156.

13. Fugger, L., Morling, N., Sandberg-Wollheim, M., *et al.* (1990) Tumor necrosis factor alpha gene polymorphism in multiple sclerosis and optic neuritis. *J. Neuroimmunol.*, **27**, 85.

14. Webb, G. C. and Chaplin, D. D. (1990) Genetic variability at the human tumor necrosis factor loci. *J. Immunol.*, **145**, 1278.

15. Messer, G., Spengler, U., Jung, M. C., *et al.* (1991) Polymorphic structure of the tumor necrosis factor (TNF) locus: an Ncol polymorphism in the first intron of the human TNF-beta gene correlates with a variant amino acid in position 26 and a reduced level of TNF-beta production. *J. Exp. Med.*, **173**, 209.

16. Nedospasov, S. A., Udalova, I. A., Kuprash, D. V., and Turetskaya, R. L. (1991) DNA sequence polymorphism at the human tumor necrosis factor (TNF) locus. Numerous TNF/lymphotoxin alleles tagged by two closely linked microsatellites in the upstream region of the lymphotoxin (TNF-beta) gene. *J. Immunol.*, **147**, 1053.

17. Jongeneel, C. V., Acha-Orbea, H., and Blankenstein, T. (1990) A polymorphic micro-satellite in the tumor necrosis factor alpha promoter identifies an allele unique to the NZW mouse strain. *J. Exp. Med.*, **171**, 2141.

18. Wilson, A. G., di Giovine, F. S., Blakemore, A. I. F., and Duff, G. W. (1992) Single base polymorphism in the human tumour necrosis factor alpha (TNF alpha) gene detectable by Ncol restriction of PCR product. *Hum. Mol. Genet.*, **1**, 353.

19. Pociot, F., Dalfonso, S., Compasso, S., Scorza, R., and Richiardi, P. M. (1995) Functional-analysis of a new polymorphism in the human TNF-alpha gene promoter. *Scand. J. Immunol.*, **42**, 501.

20. Beutler, B. and Brown, T. (1993) Polymorphism of the mouse TNF-alpha locus-sequence studies of the 3′-untranslated region and 1st intron. *Gene*, **129**, 279.

21. Waldron-Lynch, F., Adams, C., Shanahan, F., Molloy, M. G., and O'Gara, F. (1999) Genetic analysis of the 3′ untranslated region of the tumour necrosis factor shows a highly conserved region in rheumatoid arthritis affected and unaffected subjects. *J. Med. Genet.*, **36**, 214.

22. Messer, G., Spengler, U., Jung, M. C., *et al.* (1991) Allelic variation in the TNF-beta gene does not explain the low TNF-beta response in patients with primary biliary cirrhosis. *Scand. J. Immunol.*, **34**, 735.

23. Verjans, G. M., van der Linden, S. M., van Eys, G. J. J. M., *et al.* (1991). Restriction fragment length polymorphism of the tumor necrosis factor region in patients with ankylosing spondylitis. *Arthritis Rheum.*, **34**, 486.

24. Jongeneel, C. V., Briant, L., Udalova, I. A., *et al.* (1991) Extensive genetics polymorphism in the human tumor necrosis factor region and relation to extended HLA haplotypes. *Proc. Natl. Acad. Sci. USA*, **88**, 9717.

25. Wilson, A. G., de Vries, N., Pociot, F., di Giovine, F. S., van der Putte, L. B. A., and Duff, G. W. (1993) An allelic polymorphism within the human tumor necrosis factor alpha promoter region is strongly associated with HLA A1, B8 and DR3 alleles. *J. Exp. Med.*, **177**, 557.

26. Dawkins, R. L., Leaver, A., Cameron, P. U., *et al.* (1989) Some disease associated ancestral haplotypes carry a polymorphism of TNF. *Hum. Immunol.*, **26**, 91.

27. Abraham, L. J., French, M. A. H., and Dawkins, R. L. (1993) Polymorphic MHC ancestral haplotypes affect the activity of tumour necrosis factor-alpha. *Clin. Exp. Immunol.*, **92**, 14.

28. Jacob, C. O., Fronek, Z., Lewis, G. D., Koo, M., Hansen, J., and McDevitt, H. O. (1990) Heritable major histocompatibility complex II-associated differences in production of tumor necrosis factor alpha: relevance to the genetic predisposition to systemic lupus erythematosus. *Proc. Natl. Acad. Sci. USA*, **87**, 1233.

29. Badenhoop, K., Schwarz, G., Schleusener, J., *et al.* (1992) Tumor necrosis factor beta gene polymorphisms in Grave's disease. *Clin. Endocrinol. Metab.*, **74**, 287.

30. Bettinotti, M. P., Hartung, K., Deicher, H., *et al.* (1993) Polymorphism of the tumor necrosis factor beta gene in systemic lupus erythematosus: TNFB–MHC haplotypes. *Immunogenetics*, **37**, 449.

31. Wilson, A. G., Gordon, C., di Giovine, F. S., de Vries, N., van der Putte, L. B. A., Emery, P., *et al.* (1994) A genetic association between systemic lupus erythematosus and tumor necrosis factor alpha. *Eur. J. Immunol.*, **24**, 191.

32. Wilson, A. G., Clay, F. E., Crane, A. M., Cork, M. J., and Duff, G. W. (1995) Comparative genetic association of human leukocyte antigen class II and tumor necrosis factor alpha with dermatitis herpetiformis. *J. Invest. Dermatol.*, **104**, 856.

33. McManus, R., Wilson, A. G., Mansfield, J., Weir, D. G., Duff, G. W., and Kelleher, D. (1996) TNF2, a polymorphism of the tumour necrosis-alpha gene promoter, is a component of the celiac disease major histocompatibility complex haplotype. *Eur. J. Immunol.*, **26**, 2113.

34. Jacob, C. O., Hwang, F., Lewis, G. D., and Stall, A. M. (1991) Tumor necrosis factor alpha in murine systemic lupus erythematosus disease models: implications for genetic predisposition and immune regulation. *Cytokine*, **3**, 551.

35. Kroeger, K. R., Carville, K. S., and Abraham, L. J. (1997) The -308 tumor necrosis factor-alpha promoter polymorphism effects transcription. *Mol. Immunol.*, **34**, 391.

36. Wilson, A. G., Symons, J. A., McDowell, T. L., McDevitt, H. O., and Duff, G. W. (1997) Effects of a polymorphism in the human tumor necrosis factor alpha promoter on transcriptional activation. *Proc. Natl. Acad. Sci. USA*, **94**, 3195.

37. Wu, W. S. and McClain, K. L. (1997) DNA polymorphisms and mutations of the tumor necrosis factor-alpha (TNF-alpha) promoter in Langerhans cell histiocytosis (LCH). *J. Interferon Cytokine Res.*, **10**, 631.

38. Brinkman, B. M. N., Zuijdgeest, D., Kaijzel, E. L., Breedveld, F. C., and Verweij, C. L. (1996) Relevance of the tumor necrosis factor alpha (TNF alpha) -308 promoter polymorphism in TNF alpha gene regulation. *J. Inflam.*, **46**, 32.

39. Louis, E., Franchimont, D., Piron, A., *et al.* (1998) Tumour necrosis factor (TNF) gene polymorphism influences TNF-alpha production in lipopolysaccharide (LPS)-stimulated whole blood cell culture in healthy humans. *Clin. Exp. Immunol.*, **113**, 401.

40. Huang, D. R., Pirskanen, R., Matell, G., and Lefvert, A. K. (1999) Tumour necrosis factor-alpha polymorphism and secretion in myasthenia gravis. *J. NeuroImmunol.*, **94**, 165.

41. McGuire, W., Hill, A. V. S., Allsopp, C. E. M., Greenwood, B. M., and Kwiatkowski, D. (1994) Variation on the TNF-alpha promoter region associated with susceptibility to cerebral malaria. *Nature*, **371**, 508.

42. Conway, D. J., Holland, M. J., Bailey, R. L., Campbell, A. E., Mahdi, O. S. M., Jennings, R., *et al.* (1997) Scarring trachoma is associated with polymorphism in the tumor necrosis factor alpha (TNF-alpha) gene promoter and with elevated TNF-alpha levels in tear fluid. *Infect. Immun.*, **65**, 1003.

43. Carbera, M., Shaw, M. A., Sharples, C., Williams, H., Castes, M., Convit, J., *et al.* (1995) Polymorphism in tumor-necrosis-factor genes associated with mucocutaneous leishmaniasis. *J. Exp. Med.*, **182**, 1259.

44. Nadel, S., Newport, M. J., Booy, R., and Levin, M. (1996) Variation in the tumor necrosis factor-alpha gene promoter region may be associated with death from meningococcal disease. *J. Infect. Dis.*, **174**, 878.

45. Roy, S., McGuire, W., Mascie-Taylor, C. G. N., Saha, B., Hazra, S. K., Hill, A. V. S., *et al.* (1997) Tumor necrosis factor promoter polymorphism and susceptibility to lepromatous leprosy. *J. Infect. Dis.*, **176**, 530.

46. Flach, R., Majetschak, M., Heukamp, T., Jennissen, V., Flohe, S., Borgermann, J., *et al.* (1999) Relation of *ex vivo* stimulated blood cytokine synthesis to post-traumatic sepsis. *Cytokine*, **11**, 173.

47. Hurme, M. and Helminen, M. (1998) Resistance to human cytomegalovirus infection may be influenced by genetic polymorphisms of the tumour necrosis factor-alpha and interleukin-1 receptor antagonist genes. *Scand. J. Infect. Dis.*, **30**, 447.

48. Chen, G., Wilson, R., Wang, S. H., Zheng, H. Z., Walker, J. J., and McKillop, J. H. (1996) Tumour necrosis factor-alpha (TNF-alpha) gene polymorphism and expression in pre-eclampsia. *Clin. Exp. Immunol.*, **104**, 154.

49. Sullivan, K. E., Wooten, C., Schmeckpeper, B. J., Goldman, D., and Petri, M. A. (1997) A promoter polymorphism of tumor necrosis factor alpha associated with systemic lupus erythematosus in African-Americans. *Arthritis Rheum.*, **40**, 2207.

50. Fernandez-Real, J. M., Gutierrez, C., Ricart, W., Casamitjana, R., Fernandez-Castaner, M., Vendrell, J., *et al.* (1997) The TNF-alpha gene Nco1 polymorphism influences the relationship among insulin resistance, percent body fat, and increased serum leptin levels. *Diabetes*, **46**, 1468.

51. Herrmann, S. M., Ricard, S., Nicaud, V., *et al.* (1998) Polymorphisms of the tumour necrosis factor-alpha gene, coronary heart disease and obesity. *Eur. J. Clin. Invest.*, **28**, 59.

52. Demeter, J., Porzsolt, F., Ramisch, S., Schmidt, D., Schmid, M., and Messer, G. (1997) Polymorphism of the tumour necrosis factor-alpha and lymphotoxin-alpha genes in chronic lymphocytic leukaemia. *Br. J. Haematol.*, **97**, 107.

53. Chouchane, L., Ben-Ahmed, S., Baccouche, S., and Remadi, S. (1997) Polymorphism in the tumor necrosis factor-alpha promoter region and in the heat shock protein 70 genes associated with malignant tumors. *Cancer*, **80**, 1489.

54. Warzocha, K., Ribeiro, P., Bienvenu, J., Roy, P., Charlot, C., Rigal, D., *et al.* (1998) Genetic polymorphisms in the tumor necrosis factor locus influence non-Hodgkin's lymphoma outcome. *Blood*, **91**, 3574.

55. Moffatt, M. F. and Cookson, W. O. C. M. (1997) Tumour necrosis factor haplotypes and asthma. *Hum. Mol. Genet.*, **6**, 551.

56. Albuquerque, R. V., Hayden, C. M., Palmer, L. J., Laing, I. A., Rye, P. J., Gibson, N. A., *et al.* (1998) Association of polymorphisms within the tumour necrosis factor (TNF) genes and childhood asthma. *Clin. Exp. Allergy*, **28**, 578.

57. Huang, S. L., Su, C. H., and Chang, S. C. (1997) Tumor necrosis factor-alpha gene polymorphism in chronic bronchitis. *Am. J. Resp. Crit. Care Med.*, **156**, 1436.

58. Seitzer, U., Swider, C., Stuber, F., *et al.* (1997) Tumour necrosis factor alpha promoter gene polymorphism in sarcoidosis. *Cytokine*, **9**, 787.

59. Jones, D. E. J., Watt, F. E., Grove, J., Newton, J. L., Daly, A. K., Gregory, W. L., *et al.* (1999) Tumour necrosis factor-alpha promoter polymorphisms in primary biliary cirrhosis. *J. Hepatol.*, **30**, 232.

60. Bernal, W., Moloney, M., Underhill, J., and Donaldson, P. T. (1999) Association of tumor necrosis factor polymorphism with primary sclerosing cholangitis. *J. Hepatol.*, **30**, 237.

61. Zelano, G., Lino, M. M., Evoli, A., Settesoldi, D., Batocchi, A. P., Torrente, I., *et al.* (1998) Tumour necrosis factor beta gene polymorphisms in myasthenia gravis. *Eur. J. Immunogenet.*, **25**, 403.

62. Hjelmstrom, P., Peacock, C. S., Giscombe, R., Pirskanen, R., Lefvert, A. K., Blackwell, J. M., *et al.* (1998) Polymorphism in tumor necrosis factor genes associated with myasthenia gravis. *J. NeuroImmunol.*, **88**, 137.

63. Turner, D., Grant, S. C. D., Yonan, N., Sheldon, S., Dyer, P. A., Sinnott, P. J., *et al.* (1997) Cytokine gene polymorphism and heart transplant rejection. *Transplantation*, **64**, 776.

64. Frank, K. H., Fussel, M., Conrad, K., Rihs, H. P., Koch, R., Gebhardt, B., *et al.* (1998) Different distribution of HLA class II and tumor necrosis factor alleles (TNF-308.2, TNFα2 microsatellite) in anti-topoisomerase I responders among scleroderma patients with and without exposure to quartz/metal dust. *Arthritis Rheum.*, **41**, 1306.

65. Wilson, A. G., de Vries, N., van der Putte, L. B. A., and Duff, G. W. (1995) A tumor necrosis factor alpha polymorphism is not associated with rheumatoid arthritis. *Ann. Rheum. Dis.*, **54**, 601.

66. Vinasco, J., Beraun, Y., Nieto, A., Fraile, A., Mataran, L., Pareja, E., *et al.* (1997) Polymorphism at the TNF loci in rheumatoid arthritis. *Tissue Antigens*, **49**, 74.

67. Kaijzel, E. L., van Krugten, M. V., Brinkman, B. M. N., *et al.* (1998) Functional analysis of a human tumor necrosis factor alpha (TNF-alpha) promoter polymorphism related to joint damage in rheumatoid arthritis. *Mol. Med.*, **4**, 724.

68. Hohler, T., Kruger, A., Schneider, P. M., Schopf, R. E., Knop, J., Rittner, C., *et al.* (1997) A TNF-alpha promoter polymorphism is associated with juvenile onset psoriasis and psoriatic arthritis. *J. Invest. Dermatol.*, **109**, 562.

69. Arias, A. I., Giles, B., Eiermann, T. H., Sterry, W., and Pandey, J. P. (1997) Tumor necrosis factor-alpha gene polymorphism in psoriasis. *Exp. Clin. Immunogenet.*, **14**, 118.

70. Hohler, T., Kruger, A., Gerken, G., Schneider, P. M., zum Buschenfelde, K. H. M., and Rittner, C. (1998) A tumour necrosis factor-alpha (TNF-alpha) promoter polymorphism is associated with chronic hepatitis B infection. *Clin. Exp. Immunol.*, **111**, 579.

71. Hohler, T., Schaper, T., Schneider, P. M., zum Buschenfelde, K. H. M., and Marker-Hermann, E. (1998) Association of different tumor necrosis factor alpha promoter allele frequencies with ankylosing spondylitis in HLA-B27 positive individuals. *Arthritis Rheum.*, **41**, 1489.

72. Nicklin, M. J., Weith, A., and Duff, G. W. (1994) A physical map of the region encompassing the human interleukin-1 alpha, interleukin-1 beta and interleukin-1 receptor antagonist genes. *Genomics*, **19**, 382.

73. di Giovine, F. S., Takhsh, E., Blakemore, A. I. F., and Duff, G. W. (1992) Single base polymorphism in the human tumour necrosis factor alpha (TNF-α) gene detectable by NcoI restriction of PCR product. *Hum. Mol. Genet.*, **1**, 353.

74. McDowell, T. L., Symons, J. A., Ploski, R., Forre, O., and Duff, G. W. (1993) A polymorphism in the 5′ region of the interleukin-1 alpha gene is associated with juvenile chronic arthritis. *Br. J. Rheumatol.*, **32**, 162.

75. Bailly, S., di Giovine F. S., and Duff, G. W. (1993) Polymorphic tandem repeat region in interleukin-1 and alpha intron 6. *Hum. Genet.*, **91**, 85.

76. Steinkasserer, A., Koelble, K., and Sim, R. B. (1991) Length variation within intron-2 of the human IL-1 receptor antagonist protein gene (IL1RN). *Nucleic Acids Res.*, **19**, 5085.

77. Tarlow, J. K., Blakemore, A. I. F., Lennard, A., Solari, R., Hughes, H. N., Steinkasserer, A., *et al.* (1993) Polymorphism in human IL-1 receptor antagonist gene intron-2 is caused by variable numbers of an 86-bp tandem repeat. *Hum. Genet.*, **91**, 403.

78. Bailly, S., Israel, N., Fay, M., Gougerot-Pocidalo, M. A., and Duff, G. W. (1996) An intronic polymorphic repeat sequence modulates interleukin-1 alpha gene regulation. *Mol. Immunol.*, **33**, 999.

79. Pociot, F., Molvig, J., Wogensen, L., Worsaae, H., and Nerup, J. (1992) A Taq1 polymorphism in the human interleukin-1-beta (IL-1-beta) gene correlates with IL-1 beta secretion *in vitro*. *Eur. J. Clin. Invest.*, **22**, 396.

80. Lennard, A. C. (1995) Interleukin-1 receptor antagonist. *Crit. Rev. Immunol.*, **15**, 77.

81. Andus, T., Daig, R., Vogl, D., Aschenbrenner, E., Lock, G., Hollerback, S., *et al.* (1997) Imbalance of the interleukin 1 system in colonic mucosa-association with intestinal inflammation and interleukin 1 receptor antagonist genotype 2. *Gut*, **41**, 651.

82. Hurme, M. and Santtila, S. (1998) IL-1 receptor antagonist (IL-1ra) plasma levels are coordinately regulated by both IL-1ra and IL-1 beta genes. *Eur. J. Immunol.*, **28**, 2598.

83. Clay, F. E., Tarlow, J. K., Cork, M. J., Cox, A., Nicklin, M. J. H., and Duff, G. W. (1996) Novel interleukin-1 receptor antagonist exon polymorphisms and their use in allele-specific mRNA assessment. *Hum. Genet.*, **97**, 723.

84. Cox, A., Camp, N. J., Nicklin, M. J. H., di Giovine, F. S., and Duff, G. W. (1998) An analysis of linkage disequilibrium in the interleukin-1 gene cluster, using a novel grouping method for multiallelic markers. *Am. J. Hum. Genet.*, **62**, 1180.

85. Clay, F. E., Cork, M. J., Tarlow, J. K., Blakemore, A. I. F., Harrington, C. I., Lewis, F., *et al.* (1994) Interleukin-1 receptor antagonist gene polymorphism association with lichen-sclerosus. *Hum. Genet.*, **94**, 407.

86. Tarlow, J. K., Clay, F. E., Cork, M. J., Blakemore, A. I. F., McDonagh, A. J. G., Messenger, A. G., *et al.* (1994) Severity of alopecia areata is associated with a polymorphism in the interleukin-1 receptor antagonist gene. *J. Invest. Dermatol.*, **103**, 387.

87. Blakemore, A. I. F., Tarlow, J. K., Cork, M. J., Gordon, C., Emery, P., and Duff, G. W. (1994) Interleukin-1 receptor antagonist gene polymorphism as a disease severity factor in systemic lupus erythematosus. *Arthritis Rheum.*, **37**, 1380.

88. Blakemore, A. I. F., Cox, A., Gonzalez, A. M., Maskill, J. K., Hughes, M. E., Wilson, R. M., *et al.* (1996) Interleukin-1 receptor antagonist allele (IL1RN(*)2) associated with nephropathy in diabetes mellitus. *Hum. Genet.*, **97**, 369.

89. Mansfield, J. C., Holden, H., Tarlow, J. K., di Giovine, F. S., McDowell, T. L., Wilson, A. G., *et al.* (1994) Novel genetic association between ulcerative colitis and anti-inflammatory cytokine interleukin-1 receptor antagonist. *Gastroenterology*, **106**, 637.

90. Bioque, G., Crusius, J. B. A., Koutroubakis, I., Bouma, G., Kostense, P. J., Meuwissen, S. G. M., *et al.* (1995) Allelic polymorphism in IL-1 beta and IL-1 receptor antagonist. *Clin. Exp. Immunol.*, **102**, 379.

91. Brett, P. M., Yasuda, N., Yiannakou, J. Y., Herbst, F., Ellis, H. J., Vaughan, R., *et al.* (1996) Genetic and immunological markers in pouchitis. *Eur. J. Gastroenterol. Hepatol.*, **8**, 951.

92. Heresbach, D., Alizadeh, M., Dabadie, A., Le Berre, N., Colombel, J. F., Yaouanq, J., *et al.* (1997) Significance of interleukin-1 beta and interleukin-1 receptor antagonist genetic polymorphism in inflammatory bowel diseases. *Am. J. Gastroenterol.*, **92**, 1164.

93. McDowell, T. L., Symons, J. A., Ploski, R., Forre, O., and Duff, G. W. (1995) A genetic association between juvenile rheumatoid arthritis and a novel interleukin-1-alpha polymorphism. *Arthritis Rheum.*, **38**, 221.

94. Tarlow, J. K., Cork, M. J., Clay, F. E., *et al.* (1997) Association between interleukin-1 receptor antagonist (IL-1ra) gene polymorphism and early and late-onset psoriasis. *Br. J. Dermatol.*, **136**, 147.

95. Hacker, U. T., Gomolka, M., Keller, E., Eigler, A., Folwaczny, C., Fricke, H., *et al.* (1997) Lack of association between an interleukin-1 receptor antagonist gene polymorphism and ulcerative colitis. *Gut*, **40**, 623.

96. Liu, J. H., Cheng, Z. H., Yu, Y. S., Tang, Z., and Li, L. S. (1997) Interleukin-1 receptor antagonist allele: is it a genetic link between Henoch–Schonlein nephritis and IgA nephropathy? *Kidney Int.*, **51**, 1938.

97. Liu, Z. H. and Li, L. S. (1997) Polymorphism in IgA nephropathy. *Nephrology*, **3**, 63.

98. Loughrey, B. V., Maxwell, A. P., Fogarty, D. G., Middleton, D., Harron, J. C., Patterson, C. C., *et al.* (1998) An interleukin 1B allele, which correlates with a higher secretor phenotype, is associated with diabetic nephropathy. *Cytokine*, **10**, 984.

99. Pociot, F., Ronningen, K. S., Bergholdt, R., Lorenzen, T., Johannesen, J., Ye, K., *et al.* (1994) Genetic susceptibility markers in Danish patients with type-1 (insulin-dependent) diabetes—evidence for polygenicity in man. *Autoimmunity*, **19**, 169.

100. Kornman, K. S., Crane, A., Wang, H. Y., di Giovine, F. S., Newman, M. G., and Pirk, F. W. (1997) The interleukin-1 genotype as a severity factor in adult periodontal disease. *J. Clin. Periodontol.*, **24**, 72.

101. Gore, E. A., Sanders, J. J., Pandey, J. P., Palesch, Y., and Galbraith, G. M. P. (1998) Interleukin-1 beta (+3953) allele 2: association with disease status in adult periodontitis. *J. Clin. Periodontol.*, **25**, 781.

102. dela Concha, E. G., Arroyo, R., Crusius, J. B. A., Campillo, J. A., Martin, C., de Saija, E., *et al.* (1997) Combined effect of HLA-DRB1* 1501 and interleukin-1 receptor antagonist gene allele 2 in susceptibility to relapsing/remitting multiple sclerosis. *J. NeuroImmunol.*, **80**, 172.

103. Schrijver, H. M., Crusius, J. B. A., Uitdehaag, B. M. J., Gonzalez, M. A. G., Kostense, P. J., Polman, C. H., *et al.* (1999) Association of interleukin-1 beta and interleukin-1 receptor antagonist genes with disease severity in MS. *Neurology*, **52**, 595.

104. Huang, D. R., Xia, S. Q., Zhou, Y. H., Pirskanen, R., Liu, L., and Lefvert, A. K. (1998) No evidence for interleukin-4 gene conferring susceptibility to myasthenia gravis. *J. NeuroImmunol.*, **92**, 208.

105. Keen, R. W., Woodford-Richens, K. L., Lanchbury, J. S., and Spector, T. D. (1998) Allelic variation at the interleukin-1 receptor antagonist gene is associated with early post-menopausal bone loss at the spine. *Bone*, **23**, 367.

106. Perrier, S., Coussediere, C., Dubost, J. J., Albuisson, E., and Sauvezie, B. (1998) IL-1 receptor antagonist (IL-1ra) gene polymorphism in Sjogren's syndrome and rheumatoid arthritis. *Clin. Immunol. Immunopathol.*, **87**, 309.

107. Hurme, M. and Helminen, M. (1998) Polymorphism of the IL-1 gene complex in Epstein–Barr virus seronegative and seropositive adult blood donors. *Scand. J. Immunol.*, **48**, 219.
108. Francis, S. E., Holden, H., Holt, C. M., Gadsdon, P. A., and Duff, G. W. (1998) Interleukin-1 in myocardium and coronary arteries of patients with dilated cardiomyopathy. *J. Mol. Cell. Cardiol.*, **30**, 215.
109. Francis, S. E., Camp, N. J., Dewberry, R. M., *et al.* (1999) Interleukin-1 receptor antagonist gene polymorphism and coronary artery disease. *Circulation*, **99**, 861.
110. Noguchi, M., Yi, H., and Rosenblatt, H. M. (1993) Interleukin-2 receptor gamma chain mutation results in X-linked severe combined immunodeficiency in humans. *Cell*, **73**, 147.
111. Walley, A. J. and Cookson, W. O. C. M. (1996) Investigation of an interleukin-4 promoter polymorphism for associations with asthma and atopy. *J. Med. Genet.*, **33**, 689.
112. Noguchi, E., Shibasaki, M., Arinami, T., Takeda, K., Yokouchi, Y., Kawashima, T., *et al.* (1998) Association of asthma and the interleukin-4 promoter gene in Japanese. *Clin. Exp. Allergy*, **28**, 449.
113. Dizier, M. H., Sandford, A., Walley, A., Philippi, A., Cookson, W., and Demenais, F. (1999) Indication of linkage of serum IgE levels to the interleukin-4 gene and exclusion of the contribution of the (-590 C to T) interleukin-4 promoter polymorphism to IgE variation. *Genet. Epidemiol.*, **16**, 84.
114. Van den Broeck, K., Martino, G., Marrosu, M. G., *et al.* (1997) Occurrence and clinical relevance of an interleukin-4 gene polymorphism in patients with multiple sclerosis. *J. NeuroImmunol.*, **76**, 189.
115. Huang, D., Pirskanen, R., Hjelmstrom, P., and Lefvert, A. K. (1998) Polymorphisms in IL-1 beta and IL-1 receptor antagonist associated with myasthenia gravis. *J. NeuroImmunol.*, **81**, 76.
116. Hershey, G. K. K., Friedrich, M. F., Esswein, L. A., Thomas, M. L., and Chatila, T. A. (1997) The association of atopy with a gain-of-function mutation in the alpha subunit of the interleukin-4 receptor. *N. Engl. J. Med.*, **337**, 1720.
117. Kruse, S., Japha, T., Tedner, M., Sparholt, S. H., Forster, J., Kuehr, J., *et al.* (1999) The polymorphisms S503P and Q576R in the interleukin-4 receptor alpha gene are associated with atopy and influence the signal transduction. *Immunology*, **96**, 365.
118. Jacob, C. O., Mykytyn, K., and Tashman, N. (1993) DNA polymorphism in cytokine genes based on length variation in simple-sequence tandem repeats. *Immunogenetics*, **38**, 251.
119. Linker-Israeli, M., Wallace, D. J., Prehn, J. L., Nand, R., Li, L., and Klinenberg, J. R. (1996) A greater variability in the 3' flanking region of the IL-6 gene in patients with systemic lupus erythematosus (SLE). *Autoimmunity*, **23**, 199.
120. Murray, R. E., McGuigan, F., Grant, S. F. A., Reid, D. M., and Ralston, S. H. (1997) Polymorphisms of the interleukin-6 gene are associated with bone mineral density. *Bone*, **21**, 89.
121. Fishman, D., Faulds, G., Jeffery, R., Mohamed-Ali, V., Ydkin, J. S., Humphries, S., *et al.* (1998) The effect of novel polymorphisms in the interleukin-6 (IL-6) gene on IL-6 transcription and plasma IL-6 levels, and an association with systemic-onset juvenile chronic arthritis. *J. Clin. Invest.*, **102**, 1369.
122. Turner, D. M., Williams, D. M., Sankaran, D., Lazarus, M., Sinnott, P. J., and Hutchinson, I. V. (1997) An investigation of polymorphism in the interleukin-10 gene promoter. *Eur. J. Immunogenet.*, **24**, 1.

123. Eskdale, J., Wordsworth, P., Bowman, S., Field, M., and Gallagher, G. (1997) Association between polymorphisms at the human IL-10 locus and systemic lupus erythematosus. *Tissue Antigens*, **49**, 635.

124. Eskdale, J., Kube, D., Tesch, H., and Gallagher, G. (1997) Mapping of the human IL10 gene and further characterization of 5' flanking sequence. *Immunogenetics*, **46**, 120.

125. Hobbs, K., Negri, J., Klinnert, M., Rosenwasser, L. J., and Borish, L. (1998) Interleukin-10 and transforming growth factor-beta promoter polymorphisms in allergies and asthma. *Am. J. Resp. Crit. Care Med.*, **158**, 1958.

126. Mehrian, R., Quismorio, F. P., Strassmann, G., *et al.* (1998) Synergistic effect between IL-10 and bcl-2 genotypes in determining susceptibility to systemic lupus erythematosus. *Arthritis Rheum.*, **41**, 596.

127. Middleton, P. G., Taylor, P. R. A., Jackson, G., Proctor, S. J., and Dickinson, A. M. (1998) Cytokine gene polymorphisms associating with severe acute graft-versus-host disease in HLA-identical sibling transplants. *Blood*, **92**, 3943.

128. Coakley, G., Mok, C. C., Hajeer, A. H., Ollier, W. E. R., Turner, D., Sinnott, P. J., *et al.* (1998) Interleukin-10 promoter polymorphisms in rheumatoid arthritis and Felty's Syndrome. *Br. J. Rheumatol.*, **37**, 988.

129. Simmons, G., Wilkinson, D., Reeves, J. D., *et al.* (1996) Primary, syncytium-inducing human immunodeficiency virus type I isolates are dual-tropic and most can use either lestr or CCR5 as coreceptors for virus entry. *J. Virol.*, **70**, 8355.

130. Atchison, R. E., Gosling, J., Monteclaro, F. S., Franci, C., Digilio, L., Charo, I. F., *et al.* (1996) Multiple extracellular elements of CCR5 and HIV-1 entry: Dissociation from response of chemokines. *Science*, **274**, 1924.

131. Cocchi, F., De Vico, A. L., Garzino-Demo, A., Cara, A., Gallo, R. C., and Lusso, P. (1996) The V3 domain of the HIV-1 gp120 envelope glycoprotein is critical for chemokine-mediated blockade of infection. *Nature Med.*, **2**, 1244.

132. Liu, R., Paxton, W. A., Choe, S., *et al.* (1996) Homozygous defect in HIV-1 coreceptor accounts for resistance of some multiply-exposed individuals to HIV-1 infection. *Cell*, **86**, 367.

133. Huang, Y. X., Paxton, W. A., Wolinsky, S. M., *et al.* (1996) The role of a mutant CCR5 allele in HIV-1 transmission and disease progression. *Nature Med.*, **2**, 1240.

134. Cohen, O. J., Vaccarezza, N., Lam, G. K., *et al.* (1997) Heterozygosity for a defective gene for CC chemokine receptor 5 is not the sole determinant for the immunologic and virologic phenotype of HIV-infected long-term nonprogressors. *J. Clin. Invest.*, **100**, 1581.

135. Morawetz, R. A., Rizzardi, G. P., Glauser, D., *et al.* (1997) Genetic polymorphism of CCR5 gene and HIV disease: the heterozygous (CCR5/Delta ccr5) genotype is neither essential nor sufficient for protection against disease progression. *Eur. J. Immunol.*, **27**, 3223.

136. Hoffman, T. L., MacGregor, R. R., Burger, H., Mick, R., Doms, R. W., and Collman, R. G. (1997) CCR5 genotypes in sexually active couples discordant for human immuno-deficiency virus type I infection status. *J. Infect. Dis.*, **176**, 1093.

137. Ugolini, S., Moulard, M., Mondor, I., *et al.* (1997) HIV-1 gp120 induces an association between CD4 and the chemokine receptor CXCR4. *J. Immunol.*, **159**, 3000.

138. McDermott, D. H., Zimmerman, P. A., Guignard, F., Kleeberger, C. A., Leitman, S. F., and Murphy, P. M. (1998) CCR5 promoter polymorphism and HIV-1 disease progression. *Lancet*, **352**, 866.

139. Martin, M. P., Dean, M., Smith, M. W., *et al.* (1998) Genetic acceleration of AIDS progression by a promoter variant of CCR5. *Science*, **282**, 1907.

140. Winkler, C., Modi, W., Smith, M. W., *et al.* (1998) Genetic restriction of AIDS pathogenesis by an SDF-1 chemokine gene variant. *Science*, **279**, 389.
141. Garred, P., Madsen, H. O., Petersen, J., Marquart, H., Hansen, T. M., Sorensen, S. F., *et al.* (1998) CC chemokine receptor 5 polymorphism in rheumatoid arthritis. *J. Rheumatol.*, **25**, 1462.
142. Stephens, J. C., Reich, D. E., Goldstein, D. B., *et al.* (1998) Dating the origin of the CCR5-Delta 32 AIDS-resistance allele by the coalescence of haplotypes. *Am. J. Hum. Genet.*, **62**, 1507.

8 | Therapeutic manipulation of the cytokine network

FRAN BALKWILL

1. Introduction

The preceding chapters in this book describe how the cytokine network mediates cell:cell and cell:matrix interactions in normal and pathological situations. The crucial role of cytokines in disease processes is also highlighted in the chapters on viral homologues of cytokines, and on the influence of functional cytokine gene polymorphisms in immune and inflammatory responses.

Cloning of cytokine and cytokine receptor genes has allowed production of milligram quantities of purified protein for use in pre-clinical and clinical studies of acute and chronic infection, inflammatory disease, autoimmune disease, and cancer. Manipulation of the cytokine network with these recombinant proteins or other cytokine regulators provides a range of novel approaches to treating acute and chronic disease.

In a previous chapter, Brennan and Feldmann described how neutralization of key cytokines in the inflammatory cascade is of benefit in inflammatory disease. This chapter will further explore the use and mechanisms action of cytokines and cytokine antagonists in the treatment of malignancy, highlighting areas where future advances are likely.

2. Recombinant cytokine therapy of cancer

A limited number of cytokines have been manufactured in sufficient quantity for pre-clinical studies in animal models and clinical trial in malignant disease (1). Over the past ten years or so, there have been notable successes that have led to regulatory approval for clinical application. These include the following:

(a) IFN-α, either as a single subtype recombinant protein or as a mixture of human cell source subtypes, in some haematological malignancies and as adjuvant therapy in high-risk melanoma (2, 3).

(b) IL-2 in a minority of patients with malignant melanoma and renal cell carcinoma (4, 5).

(c) TNF-α in destroying tumour vasculature and inducing complete tumour regression when given loco-regionally (6–8).

(d) G- and GM-CSF used to accelerate bone marrow recovery, reduce infections after myeloablative therapies, and to stimulate peripheral blood stem cell mobilization (9–11).

(e) IL-11 in the treatment of thrombocytopenia following high dose chemotherapy (12).

In addition, IFN-β has been approved for the treatment of multiple sclerosis (13), IFN-α for some viral diseases especially chronic active hepatitis (14), and IFN-γ for the treatment of chronic granulomatous disease (15).

The major problems encountered with cytokine therapies are their rapid clearance from the circulation (16, 17), the antigenicity of the recombinant protein (18) and, sometimes, severe and dose-limiting toxicity (19). Most cytokines so far clinically tested are capable of perturbing the endogenous cytokine network and inducing symptoms that range from a mild flu-like syndrome to the signs and symptoms of endotoxic shock. However, with extensive trial of dose and schedule, it is possible to achieve useful clinical outcome with acceptable toxicity, using optimal biological doses that are generally lower than the maximally tolerated dose (20, 21).

There are also some cytokines, e.g. IL-3, IL-4, IL-5, IL-6, IL-8, IL-10, and IL-12, that have shown promise in pre-clinical studies but have yet to show clinical utility. Reasons for a low correlation between pre-clinical studies and clinical trial include:

(a) Optimistic interpretation of animal model and tissue culture data.

(b) Incomplete understanding of endogenous cytokines in the microenvironment of the disease.

(c) Inappropriate methods of administration.

(d) Dose-limiting toxicity.

2.1 IFN-α therapy of malignant disease

IFN-α is the first human therapeutic protein to influence the survival of cancer patients. Given as a single agent, it is clinically useful in the following malignancies:

- hairy cell leukaemia (22, 23)
- Kaposi's sarcoma (24)
- chronic myelogenous leukaemia (23, 25)
- B and T cell lymphoma (26)
- myeloma (27, 28)
- renal cell carcinoma (29, 30)
- melanoma (31, 32)

For instance, in randomized Phase III studies of patients with previously untreated clinically aggressive follicular lymphoma, IFN-α administration is associated with prolonged time to treatment failure and, in two studies, a survival benefit (26). IFN-α maintenance also prolongs time to treatment failure in follicular lymphoma patients with high tumour burden and minimal residual disease after dose-intensive cytoreductive chemotherapy. IFN-α is the best available conservative treatment for early stage chronic myelogenous leukaemia and is considered a safe and effective alternative to allogeneic bone marrow transplantation (23, 25). Sustained responses occur in a majority of patients (> 75%) and in addition to reducing leukaemia cell mass, there is a gradual reduction in the frequency of cells bearing the underlying 9–22 chromosomal translocation to a level below that detectable by RT–PCR (25). In patients with high-risk resected cutaneous melanoma, IFN-α treatment for 48 weeks significantly prolonged relapse-free and overall survival when compared with observation alone (32). Many agents have been subjected to randomized control trial in this difficult disease but IFN-α is the first to show a significant benefit.

As a general rule, IFN-α works best where there is minimal tumour burden, but a mechanism of action is not entirely clear. The following are suggested from *in vitro* studies, experimental animal experiments, and responses in clinical trial.

(a) Direct inhibition of tumour cell growth (33).

(b) Action as a prototype tumour suppresser protein that represses the malignant phenotype in some cancers that are capable of differentiation (34).

(c) Down-regulation of responses to autocrine growth/survival factors (35).

(d) Induction of tumour-suppresser proteins such as the protein kinase that phosphorylates the eukaryotic peptide chain initiation factor; the IFN regulatory proteins IRF-1 and IRF-2, and a latent endoribonuclease RNase L (34).

(e) Inhibition of angiogenesis (34, 36).

(f) Inhibition of tumour cell signal transduction pathways (37).

In addition, some tumour cells may be deficient in IFN-α production (38). This deficiency may lie in the loss of genes in p21–22 from chromosome 9, where the IFN gene cluster lies, and may be secondary to the loss of the nearby retinoic acid receptor (39). However, in cell lines derived from some malignant melanomas, a deficiency in IFN-α secretion was caused by disruption of a *trans*-acting IFN-α gene transcription factor (40).

In spite of the large number of potential mechanisms of action, it is possible that optimal activity will be achieved by combining IFN-α with other tumour suppressers such as retinoids, or specific antagonists of autocrine growth/survival factors, such as TNF-α (see Section 4.1).

2.2 IFN-γ therapy of malignant disease

As the activities of IFN-γ were discovered, its potential as an anti-cancer agent seemed great. This cytokine is a potent growth inhibitory molecule with a major role in antigen presentation and other activating and regulatory roles in the Th1 immune response (41, 42). Moreover, as detailed below, induction of IFN-γ in the IL-12 treated tumour microenvironment, is central to its potent anti-tumour actions (43). However, for reasons that are not entirely clear, IFN-γ has not, in general, shown clinical effectiveness in malignancy, although it is useful in the treatment of the rare immunodeficiency syndrome, chronic granulomatous disease (15). The reason for this lack of activity may be pharmacokinetic or pharmacodynamic. Experimental animal studies have clearly shown the importance of IFN-γ induction in the tumour microenvironment during IL-12 therapy (43–45) but the same tumours do not respond to systemic IFN-γ (44, 45). The cytokine is rapidly cleared and the IFN-γ receptor is expressed widely.

The most promising clinical study used intraperitoneal administration of relatively high doses of IFN-γ in patients with minimal residual disease after chemotherapy for ovarian cancer (46). Parallel studies in nude mouse ovarian cancer xenograft models provided evidence for potent anti-proliferative and apoptotic actions at clinically achievable doses (47). *In vitro* experiments in ovarian tumour cell lines and freshly isolated tumour cells have confirmed that approximately three-quarters of patients are likely to be sensitive to these actions of IFN-γ even after several cycles of chemotherapy and relapse of disease. Sustained exposure to IFN-γ for three days *in vitro* or seven days in the animal models, is crucial to response and is associated with sustained increases in the cyclin kinase inhibitor p21 and the transcription factor IRF-1. Further clinical trial of IFN-γ may be warranted in ovarian cancer, possibly in combination with an antagonist of cytokine signalling (see Section 3.1).

2.3 TNF-α therapy of malignant disease

About one hundred years ago, William Coley treated a series of advanced cancer patients with filtrates of bacterial cultures (48, 49). There were undoubtedly some dramatic and well-documented cures in a minority of patients, an effect now attributed to hyperthermia and the local or systemic induction of cytokines, particularly TNF-α. A further 70 years of laboratory experiments led to the isolation and subsequent cloning of this cytokine.

The anti-tumour actions of TNF-α have been most extensively studied in the murine Meth-A sarcoma model (50). Meth-A tumour cells were resistant to the cytotoxic effect of TNF *in vitro*, yet systemic administration of TNF-α consistently caused haemorrhagic necrosis of subcutaneous (vascular) but not intraperitoneal (avascular, ascitic) tumours (50–52). Histologically, tumour blood vessels were occluded by thrombus, and tumour vasculature selectively killed. Similar effects, with either TNF-α alone or in combination with IFN-γ, were seen in tumour xenograft models.

When the recombinant cytokine was used, however, in a Phase I clinical trial, tumour response was minimal (53, 54). The maximum tolerated dose (MTD) for recombinant TNF-α was 200 mg/m^2 day with hypotension being the principal limiting factor. Other acute side-effects included fever, rigors, chills, headache, pulmonary toxicity, chest pain, and intravascular coagulation (53–55). Responses to TNF therapy given at relatively non-toxic doses were also extremely rare.

However, the ability of this cytokine to induce tumour necrosis has recently been confirmed in one specific clinical setting. High doses of loco-regional TNF-α (four times the MTD) given with local chemotherapy, and combined in some studies with systemic IFN-γ, mediated the specific destruction of tumour vasculature (56, 57). Response rates of 50–90% were achieved in patients with melanoma, squamous cell carcinoma, and soft tissue sarcoma, whose tumours were accessible to this isolated limb perfusion (ILP) approach. The impressive anti-tumour effects were associated with vascular-mediated damage although the local anti-tumour effects were not accompanied by regression of distant metastasis.

Recent study of the action of TNF-α and IFN-γ combinations on human endothelial cells *in vitro* and *in vivo* suggests they reduce αVβ3 integrin activation. This integrin is normally expressed on angiogenic endothelial cells in physiological or pathological situations and inhibition results in endothelial cell apoptosis and disruption of neovasculature (6).

2.4 IL-12 therapy of malignant disease

As described in Chapter 4, IL-12 is a heterodimeric cytokine, produced mainly by macrophages/monocytes (58–61). It induces a range of cytokines and promotes the establishment of a Th1-type response. IL-12 also induces regression and cure of a range of syngeneic transplantable tumours, with local induction of IFN-γ playing a major role in its action (43, 44, 62, 63). Treatment with IL-12 promotes a cytotoxic anti-tumour immune response and, most recently, anti-angiogenic effects have been described (45, 64). These actions of IL-12 can occur together in a single model as we have shown in a novel transplantable murine breast cancer, HTH-K (44, 45). The anti-angiogenic actions of IL-12 are thought to be largely caused by local IFN-γ production and its subsequent induction of the angiostatic chemokine IP-10 (43, 65). However, in the HTH-K model, IL-12 also regulated production of the angiogenic cytokine VEGF as well as the matrix metalloprotease MMP-9 and its natural inhibitor TIMP-1 (45). IFN-γ reduced tumour cell production of VEGF *in vitro*, suggesting that IL-12-induced IFN-γ may be responsible for the decline in VEGF levels *in vivo*. There was also evidence that IL-12 regulated stromal cell interactions leading to MMP-9 inhibition and TIMP-1 production. Thus, at least three mechanisms may be involved in IL-12 regulation of angiogenesis: induction of an angiostatic chemokine, IP-10; removal of a pro-angiogenic stimulus, and blocking the release and activity of MMPs. One logical extension of these results would be to use IL-12 and an MMP inhibitor sequentially in a therapeutic setting.

Pre-clinical experiments with IL-12 are notable for three main reasons; first the impressive responses even with established tumours; secondly, the very detailed information on tumour microenvironment during therapy; and thirdly, the multiple actions of IL-12 in any one tumour model. The multiple actions are summarized in Fig. 1 using the HTH-K model as an example.

These promising results have not, as yet, translated into a useful cancer therapy.

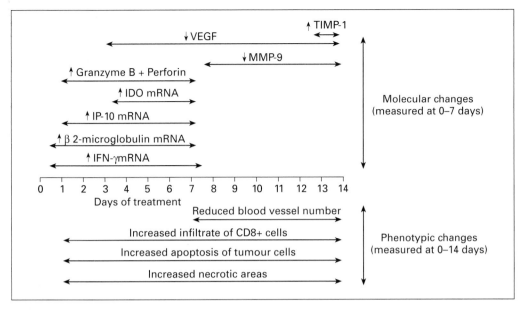

Fig. 1 A summary of the multiple effects of IL-12 in the microenvironment of a transplantable murine breast tumour HTH-K (based on work described in refs 44 and 45).

Unexpectedly high toxicity was a particular problem of the first trials (66), but more recent Phase I studies in patients with advanced cancer have shown that lower doses can have marked effects on immune parameters and some anti-tumour activity has been recorded (67, 68). The most severe toxicities occurred with the first injection and were milder upon further treatment (68). Phase II studies in previously untreated patients are now underway.

3. The potential of cytokine antagonist therapy in cancer

There are many ways (e.g. anti-cytokine antibodies, signal transduction inhibitors, and soluble receptors) to target cytokine signalling. Two inhibitors of TNF-α signalling, a humanized antibody and a dimeric receptor antagonist, have been well tolerated in clinical trial, and, in contrast to cytokine therapies, are characterized by stable and long-term circulating levels (1, 69). Soluble receptors to IL-1 and IL-4 are currently undergoing human safety trials as cytokine antagonists.

Approval for clinical use has recently been given for the antibodies to TNF-α in the treatment of fistula in Crohn's disease and the soluble dimeric TNF-α receptors in the treatment of rheumatoid arthritis (1). These TNF-α antagonists are likely to be useful in a number of other inflammatory diseases and there is a case for their use in the management of malignant disease.

3.1 Inflammatory cytokines and cancer—the case for cytokine antagonist therapy

The well-established association between chronic inflammatory diseases such as ulcerative colitis and hepatitis, and increased risk of cancer (70, 71) may be linked to chronic production of cytokines such as TNF-α and IL-1α/β. Experimental evidence linking inflammatory cytokines to cancer development and progression includes the following:

(a) TNF-α can act as an endogenous tumour promoter in *in vitro* assays (72).

(b) TNF-α production is down-regulated by cancer preventative agents such as caventol and tamoxifen (73).

(c) Pentoxifylline, an inhibitor of pro-inflammatory cytokine synthesis, reduces cutaneous inflammation and carcinogen/promoter-induced papilloma formation in mice (74).

(d) Pre-treatment of experimental tumour cells with TNF-α enhances their lung colonizing capacity (75).

(e) Overexpression of TNF-α renders cells invasive in a peritoneal model (76).

(f) Mice deficient in TNF-α are resistant to carcinogenesis (77).

(g) Pre-treatment of mice with TNF-α promotes development of experimental liver metastases (78).

(h) Pre-treatment of mice with the IL-1 receptor antagonist inhibits organ colonization of metastatic tumour cells (79, 80).

Aside from this experimental evidence, TNF-α is also implicated in the pathophysiology of some human cancers. For instance the cytokine is expressed in epithelial tumour islands of human ovarian cancer, where it may play a role in tumour/stroma interactions, the level of TNF-α expression increasing with severity of disease (81). In ovarian tumour biopsies, TNF-α co-localizes with MMP-9, in tumour-associated macrophages (82), and the combination of autocrine TNF-α and a soluble tumour cell-derived factor up-regulates production of this MMP (83). TNF-α also induces tumour cell production of the chemokine MCP-1, which is produced in the ovarian tumour microenvironment, and may control the extent and distribution of the host cell infiltrate (84, 85). Further evidence for the involvement of TNF-α in ovarian malignancies comes from human tumour xenograft models where TNF-α injection converts ascitic free-floating tumour to solid tumours with well-developed stroma (86). These actions of TNF-α in the tumour microenvironment of ovarian cancer are summarized in Fig. 2.

A role for TNF-α in promoting development of tumour stroma and controlling host/tumour interactions, may be analogous to its roles in inflammatory disease. As detailed above, high doses of TNF-α delivered locally to the tumour site, cause disruption of the tumour vasculature followed by tumour necrosis (6). These destructive actions of TNF can also be seen in some acute inflammatory situations. However, lower doses of endogenous TNF-α can promote tissue repair via stimulation of fibroblasts and neovasculature (87–89). The actions of TNF-α in the tumour

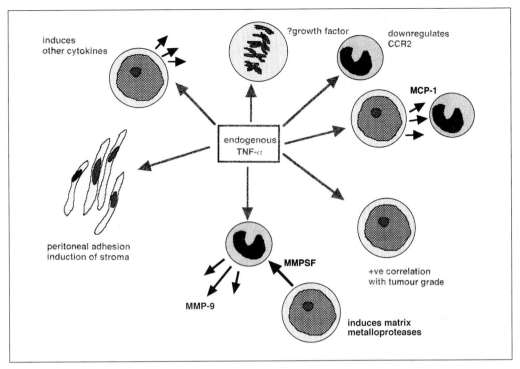

Fig. 2 Potential actions of the cytokine TNF-α in the microenvironment of epithelial ovarian cancer (based on work described in refs 80–85).

microenvironment may be similar, except that TNF-α produced chronically in the tumour microenvironment does not lead to resolution of the lesion.

3.2 Other cytokine signalling targets

Other malignancies in which cytokine signalling could be targeted include lymphoma, where TNF-α is once again implicated and where the presence or two or more high producer alleles in the TNF/LT locus is associated with poor prognosis (90), multiple myeloma where IL-6 is a growth factor for malignant plasma cells and their precursors (91), and squamous cell carcinoma where there is tumour cell expression of cytokines such as IL-1α, IL-β, and IL-6 (92, 93). In any of these conditions the pharmacokinetic superiority of cytokine antagonists over rapidly cleared cytokine proteins, may be of significance.

4. Mechanisms of action of cytokine therapy in malignancy

There appear to be two well-defined responses of experimental and human tumours to cytokine therapy. The first, typified by IFN-α sensitive haematological tumours, is a slow differentiation or destruction of the malignant clone after several months or years of daily or thrice weekly therapy. Chronic exposure to cytokine seems to be more successful than higher doses given as single injections. Response to IFN-α may also be achieved in a maintenance setting after chemotherapy and/or surgery to reduce the tumour burden. In successful protocols, there is little suggestion of a host immune response to tumour, although there is evidence for this in animal models of syngeneic transplantable tumours.

It is not clear whether this slow response will be seen with other tumour types or cytokine therapies. IFNs α, β, and γ, are unusual cytokines in that they are not known to be mitogens. Moreover, there is increasing molecular evidence that they, or the genes they induce, are tumour suppressers and that there are defects in the IFN system in several tumour types. Thus IFNs, as growth suppressing cytokines, may be deregulated during the evolution of malignancy. Any novel cytokine that is characterized as a growth suppresser, especially in epithelial cell, may have similar potential.

The second well-characterized response to a cytokine, most likely with IL-2, 12, and TNF-α, is a more rapid regression of established tumour caused by specific destruction of tumour vasculature and/or cytotoxic lymphocyte infiltration. This response may lead to specific immunity and long-term survival in a minority of experimental animals and patients, but is often associated with severe side-effects that resemble endotoxic shock. For an individual cytokine, mechanisms of action are likely to be complex with multiple cytokine actions on tumour and stroma contributing. It is unlikely that one single action of a cytokine will be sufficient for an anti-tumour effect.

5. Future directions

5.1 Local delivery

The rapid degradation or elimination of cytokines delivered systemically can decrease their effectiveness. An alternative is local delivery of cytokine gene or protein at the primary tumour site. Both these approaches could resemble cytokine release *in vivo* and target the anti-tumour response with minimal side-effects. As we understand more of the local cytokine environment of the tumours and the local actions of cytokines, methods for targeting cytokines to the tumour site becomes of prime importance.

5.1.1 Gene delivery

A range of cytokines have been tested pre-clinically in gene delivery strategies. In some experiments retroviral vectors have been used to deliver the cytokine, but the most common technique is to transfect tumour cells with cytokines *ex vivo*. Cytokines that have been tested include IFN-α, IFN-γ, and IL-2. IL-2 gene delivery has been compared with several other cytokines (IFN-γ, IFN-α, IL-7, granulocyte-macrophage colony-stimulating factor; GM-CSF). IL-2 was as good as, or superior to, other cytokines in mediating local anti-tumour effects, especially in nude mice.

Tumour cells engineered to produce IL-4 by transfection also had reduced or absent tumourgenicity *in vivo*, and they inhibited growth of untransfected cells when they were mixed with these before injection into animals (94, 95). With IL-12, one approach has been to transfect fibroblasts with IL-12 and then mix these with tumour cells.

Genes for IL-6, TNF-α, IL-2, and IFN-γ have been directly transfected into tumours *in vivo* by particle-mediated gene transfer (96). This method of gene transfer resulted in significant serum levels of IL-6 and IFN-γ and transgene expression was detected in treated tissues.

An increase in the local concentration of cytokines caused by gene delivery seems to have three main effects (97).

(a) A marked local inflammatory response, with the type of effector cell being dependent on the cytokine gene delivered.

(b) Tumour growth inhibition and sometimes regression.

(c) Long-lasting immunity to tumour rechallenge.

Prompted by encouraging pre-clinical studies, there are a range of clinical trials, planned or underway, using cytokine-directed gene therapy. Several approaches are being followed, all with the aim of provoking powerful local inflammation and generating systemic and specific immune responses to the tumour. The cytokines IL-2, IL-4, GM-CSF, and IL-12 are most frequently used in such protocols. Proposals include transfection of the relevant gene into autologous or allogeneic tumour cells or stromal components *ex vivo*. Other trials use cell-specific promoters to direct the gene *in vivo*, with either direct injection, viral or liposomal vectors being used for delivery.

In a recently reported Phase I study of IL-12 gene delivery, melanoma cells were expanded from surgically removed metastasis, transfected with cytokine by ballistic gene transfer, irradiated, and then injected subcutaneously (96). Although there were some changes in immunological parameters and lymphocyte infiltrates were seen in tumours, only one minor clinical response was recorded.

5.1.2 Protein delivery

Gene therapy is not the only option for localizing cytokines. Slow release delivery systems for cytokine protein may also be effective and provoke useful biological responses, as a study with GM-CSF microspheres showed (98). High doses of GM-CSF were incorporated into gelatin/chondroitin sulfate microspheres and mixed with B16/F10 tumour cells prior to s.c. injection. This mixture produced an intense host inflammatory infiltrate consisting of numerous eosinophils and macrophages. By 12 days post-injection of the GM-CSF neither microspheres or tumour cells were detected but a moderate infiltrate persisted. This was reflected in an increase in mouse survival.

5.2 The importance of understanding the tumour cytokine network

Cytokine therapy is given against a background of a complex and highly regulated endogenous cytokine network, not only in the whole organism, but in the tumour microenvironment. Endogenous cytokines are important mediators of tumour/stroma communication, as well as being autocrine and paracrine regulators of tumour cell growth and survival. However, the influence of these endogenous cytokines is generally not considered when designing therapies. Defining the cytokine context of tumours may not only enhance our understanding of tumour biology, but may lead to novel and more specific cytokine therapies.

One of the tumours in which this network is being examined is human epithelial ovarian cancer (99, 100). This tumour microenvironment is rich in mRNA for growth factors, pro-inflammatory cytokines, and chemokines, but weak in lymphocyte-associated cytokines (47, 101). The cytokine network has a great deal of redundancy and it is possible that targeting a single endogenous cytokine would be ineffective. However, a combination of cytokines and inhibitors, delivered systemically or locally, could provide a greater degree of specificity whether the desired result is apoptotic tumour cell death, destruction of the tumour vasculature, induction of a tumour specific host immune response or, optimally, a combination of all three.

5.3 Sequential or combination biological therapy

Major advances in chemotherapy of cancer came from a combination of agents that target different aspects of tumour cell metabolism. Although a combination of cytokines has been associated with increased and often unacceptable toxicity, sequential application of cytokines with different actions, combined with novel drugs that are likely to complement cytokine action, is an attractive option. Combination of IFN-α with other anti-angiogenic drugs is a possibility or, as detailed above, sequential administration of cytokines and agents such as matrix metalloprotease inhibitors, that disrupt tumour/stroma communication. Cocktails of down-regulatory cytokines and specific cytokine antagonists may have less toxicity. If designed with a knowledge of the growth/survival factors that typify a certain tumour, these may, over a short period of exposure, induce apoptosis of genetically damaged tumour cells, without affecting their normal counterparts. Such combinations may also promote anti-tumour responses from host cells.

6. Summary

It is now more than twenty years since cytokines were first used as cancer treatments. There is no doubt that some cytokines have proved useful and cytokine antagonists are likely to provide successful treatments in the next few years. Further exploitation of the cytokine network in malignancy and other disease states will come from understanding of the cytokine/receptor systems that regulate individual diseases;

development of small molecule inhibitors of extracellular cytokine binding and intracellular signalling, and carefully designed combinations of the above.

References

1. Gillis, S. and Williams, D. E. (1998) Cytokine therapy: lessons learned and future challenges. *Curr. Opin. Immunol.*, **10**, 501.
2. Gutterman, J. U. (1994) Cytokine therapeutics: lessons from interferon α. *Proc. Natl. Acad. Sci. USA*, **91**, 1198.
3. Pfeffer, L. M., Dinarello, C. A., Herberman, R. B., *et al.* (1998) Biological properties of recombinant α-interferons: 40th anniversary of the discovery of interferons. *Cancer Res.*, **58**, 2489.
4. Caligiuri, M. A., Murray, C., Soiffer, R. J., *et al.* (1991) Extended continuous infusion low-dose recombinant interleukin-2 in advanced cancer: prolonged immunomodulation without significant toxicity. *J. Clin. Oncol.*, **9**, 2110.
5. Mittelman, A., Puccio, C., Ahmed, T., Zeffren, J., Choudhury, A., and Arlin, Z. (1991) A Phase II Trial of Interleukin-2 by continuous infusion and interferon by intramuscular injection in patients with renal call cancer. *Cancer*, **68**, 1699.
6. Ruegg, C., Yilmaz, A., Bieler, G., Bamat, J., Chaubert, P., and Lejeune, F. J. (1998) Evidence for the involvement of endothelial cell integrin aVb3 in the disruption of the tumor vasculature induced by TNF and IFN-γ. *Nature Med.*, **4**, 408.
7. Eggermont, A. M. M., Koops, H. S., Lienard, D., Kroon, B. B. R., van Geel, A. N., Hoekstra, H. J., *et al.* (1996) Isolated limb perfusion with high-dose tumor necrosis factor-α in combination with interferon-γ and melphalan for nonresectable extremity soft tissue sarcomas: a multicenter trial. *J. Clin. Oncol.*, **14**, 2653.
8. Lienard, D., Ewalenko, P., Delmotte, J.-J., Renard, N., and Lejeune, F. J. (1992) High dose recombinant tumour necrosis factor alpha in combination with melphalan and interferon gamma in isolation perfusion of the limbs in melanoma and sarcoma. *J. Clin. Oncol.*, **10**, 52.
9. Ozzello, L., Habif, D. V., De Rosa, C. M., and Cantell, K. (1992) Cellular events accompanying regression of skin recurrences of breast carcinomas treated with intralesional injections of natural interferons alpha and gamma. *Cancer Res.*, **52**, 4571.
10. Whyte, M. and Evan, G. (1995) Apoptosis. The last cut is the deepest. *Nature*, **376**, 17.
11. Nicholson, D. W., Ali, A., Thornberry, N. A., *et al.* (1995) Identification and inhibition of the ICE/CED-3 protease necessary for mammalian apoptosis. *Nature*, **376**, 37.
12. Du, X. X. and Williams, D. A. (1994) Interleukin-11: a multifunctional growth factor derived from the hematopoietic microenvironment. *Blood*, **83**, 2023.
13. Bonn, D. (1998) Trial of interferon-β in multiple sclerosis stopped early. *Lancet*, **351**, 573.
14. Terrault, N. and Wright, T. (1995) Interferon and hepatitis C. *N. Engl. J. Med.*, **332**, 1509.
15. Dana-Farber Ca Inst. (1991) Interferon-γ and chronic granulomatous disease. *Curr. Opin. Immunol.*, **3**, 61.
16. Piscitelli, S. C., Reiss, W. G., Figg, W. D., and Petros, W. P. (1997) Pharmacokinetic studies with recombinant cytokines. Scientific issues and practical considerations. *Clin. Pharmacokinet.*, **32**, 368.
17. Wills, R. J. (1990) Clinical pharmacokinetics of interferons. *Clin. Pharmacokinet.*, **19**, 390.
18. Schiemann, W. P., Graves, L. M., Baumann, H., Morella, K. K., Gearing, D. P., Nielsen, M. D., *et al.* (1995) Phosphorylation of the human leukemia inhibitory factor (LIF) receptor by

mitogen-activated protein kinase and the regulation of LIF receptor function by heterologous receptor activation. *Proc. Natl. Acad. Sci. USA*, **92**, 5361.

19. Borden, E. C. and Parkinson, D. (1998) A perspective on the clinical effectiveness and tolerance of interferon-α. *Semin. Oncol.*, **25**, 3.

20. Kovacs, J. A., Baseler, M., Dewar, R. J., *et al.* (1995) Increases in CD4 T lymphocytes with intermittent courses of interleukin-2 in patients with human immunodeficiency virus infection. A preliminary study. *N. Engl. J. Med.*, **332**, 567.

21. Borden, E. C. and Wadler, S. (1996) Interferons as biochemical modulators. *J. Clin. Oncol.*, **14**, 2627.

22. Aderka, D., Michalevicz, R., Daniel, Y., *et al.* (1988) Recombinant interferon alpha-C for advanced hairy cell leukemia. *Cancer*, **61**, 2207.

23. Moriuchi, H., Moriuchi, M., Combadiere, C., Murphy, P. M., and Fauci, A. S. (1996) CD8+ T-cell-derived soluble factor(s), but not β-chemokines, RANTES, MIP-1α, and MIP-1β, suppress HIV-1 replication in monocyte/macrophages. *Proc. Natl. Acad. Sci. USA*, **93**, 15341.

24. Krown, S. E., Real, F. X., Cunningham-Rundles, S., Myskowski, P. L., Koziner, B., Fein, S., *et al.* (1983) Preliminary observations on the effect of leucocyte A interferon in homosexual men with Kaposi's sarcoma. *N. Engl. J. Med.*, **308**, 1071.

25. Kurzrock, R., Estrov, Z., Kantarjian, H., and Talpaz, M. (1998) Conversion of interferon-induced, long-term cytogenetic remissions in chronic myelogenous leukemia to polymerase chain reaction negativity. *J. Clin. Oncol.*, **16**, 1526.

26. Ozer, H., Wiernik, P. H., Giles, F., and Tendler, C. (1998) Recombinant interferon-α therapy in patients with follicular lymphoma. *Cancer*, **82**, 1821.

27. Peest, D. (1996) Cytokine therapy in multiple myeloma. *Br. J. Haematol.*, **94**, 425.

28. Browman, G. P., Bergsagel, D., Sicheri, D., *et al.* (1995) Randomized trial of interferon maintenance in multiple myeloma: a study of the National Cancer Institute of Canada Clinical Trials Group. *J. Clin. Oncol.*, **13**, 2354.

29. Figlin, R. A., Dekernion, J. B., Mukamel, E., Palleroni, A. V., Itri, L. M., and Sarna, G. P. (1988) Recombinant interferon alfa-2A in metastatic renal cell carcinoma: assessment of antitumour activity and anti-interferon antibody formation. *J. Clin. Oncol.*, **6**, 1604.

30. Krown, S. E. (1987) Interferon treatment of renal cell carcinoma: current status and future prospects. *Cancer*, **59**, 647.

31. Creagan, E. T., Dalton, R. J., Ahmann, D. L., Jung, S.-H., Morton, R. F., Langdon, R. M. Jr., *et al.* (1995) Randomized, surgical adjuvant clinical trial of recombinant interferon alfa-2a in selected patients with malignant melanoma. *J. Clin. Oncol.*, **13**, 2776.

32. Kirkwood, J. M., Strawderman, M. H., Ernstoff, M. S., Smith, T. J., Borden, E. C., and Blum, R. H. (1996) Interferon Alfa-2b adjuvant therapy of high-risk resected cutaneous melanoma: the Eastern Cooperative Oncology Group Trial EST 1684. *J. Clin. Oncol.*, **14**, 7.

33. Lindner, D. J., Kalvakolanu, D. V., and Borden, E. C. (1997) Increasing effectiveness of interferon-α for malignancies. *Semin. Oncol.*, **24**, S9.99.

34. Pectasides, D., Kayianni, H., Facou, A., Bobotas, N., Barbounis, V., Zis, J., *et al.* (1991) Correlation of abdominal computed tomography scanning and second-look operation findings in ovarian cancer patients. *Am. J. Clin. Oncol.*, **14**, 457.

35. Kubin, M., Chow, J. M., and Trinchieri, G. (1994) Differential regulation of interleukin-12 (IL-12), tumor necrosis factor α, and IL-1β production in human myeloid leukemia cell lines and peripheral blood mononuclear cells. *Blood*, **83**, 1847.

36. Dinney, C. P. N., Bielenberg, D. R., Perrotte, P., Reich, R., Eve, B. Y., Bucana, C. D., *et al.* (1997) Inhibition of basic fibroblast growth factor expression, angiogenesis, and growth of

human bladder carcinoma in mice by systemic interferon-α administration. *Cancer Res.*, **58**, 808.

37. Colotta, F., Sciacca, F. L., Sironi, M., Luini, W., Rabiet, M. J., and Mantovani, A. (1994) Expression of monocyte chemotactic protein-1 by monocytes and endothelial cells exposed to thrombin. *Am. J. Pathol.*, **144**, 975.

38. Corey, S. J., Burkhardt, A. L., Bolen, J. B., Geahlen, R. L., Tkatch, L. S., and Tweardy, D. J. (1994) Granulocyte colony-stimulating factor receptor signaling involves the formation of a three-component complex with Lyn and Syk protein–tyrosine kinases. *Proc. Natl. Acad. Sci. USA*, **91**, 4683.

39. Messing, A., Chen, H. Y., Palmiter, R. D., and Brinster, R. L. (1985) Peripheral neuropathies, hepatocellular carcinomas and islet cell adenomas in transgenic mice. *Nature*, **316**, 461.

40. Vaglini, M., Belli, F., Ammatuna, M., *et al.* (1994) Treatment of primary or relapsing limb cancer by isolation perfusion with high-dose alpha-tumor necrosis factor, gamma-interferon, and melphalan. *Cancer*, **73**, 483.

41. Billiau, A., Heremans, H., Vermeire, K., and Matthys, P. (1998) Immunomodulatory properties of interferon-gamma. An update. *Ann. NY Acad. Sci.*, **856**, 22.

42. Pestka, S., Kotenko, S. V., Muthukumaran, G., Izotova, L. S., Cook, J. R., and Garotta, G. (1997) The interferon γ (IFN-γ) receptor: a paradigm for the multichain cytokine receptor. *Cytokine Growth Factor Rev.*, **8**, 189.

43. Brunda, M. J., Luistro, L., Hendrazak, J. A., Fontoulakis, M., Garotta, G., and Gately, M. K. (1995) Role of interferon-gamma in mediating the antitumor efficacy of interleukin-12. *J. Immunother.*, **17**, 71.

44. Dias, S., Thomas, H., and Balkwill, F. (1998) Multiple molecular and cellular changes associated with tumour stasis and regression during IL-12 therapy of a murine breast cancer model. *Int. J. Cancer*, **75**, 151.

45. Dias, S., Boyd, R., and Balkwill, F. (1998) IL-12 regulates VEGF and MMPs in a murine breast cancer model. *Int. J. Cancer*, **78**, 361.

46. Screpanti, I., Musiani, P., Bellavia, D., *et al.* (1996) Inactivation of the IL-6 gene prevents development of multicentric Castleman's disease in C/EBPβ-deficient mice. *J. Exp. Med.*, **184**, 1561.

47. Burke, F., Relf, M., Negus, R., and Balkwill, F. (1996) A cytokine profile of normal and malignant ovary. *Cytokine*, **8**, 578.

48. Coley, W. B. (1891) Contribution to the knowledge of sarcoma. *Ann. Surg.*, **14**, 199.

49. Coley Nauts, H., Fowler, G. A., and Bogatko, F. H. (1953) A review of the influence of bacterial infection and of bacterial products (Coley's toxins) on malignant tumors in man. *Acta Med. Scand.*, 29.

50. Brunda, M. J., Luistro, L., Rumennik, L., *et al.* (1996) Antitumor activity of interleukin 12 in pre-clinical models. *Cancer Chemother. Pharmacol.*, **38**, S16.

51. Baserga, R. and Rubin, R. (1993) Cell cycle and growth control. *Crit. Rev. Eukaryot. Gene Expr.*, **3**, 47.

52. Nikkari, S. T., Geary, R. L., Hatsukami, T., Ferguson, M., Forough, R., Alpers, C. E., *et al.* (1996) Expression of collagen, interstitial collagenase, and tissue inhibitor of metalloproteinases-1 in restenosis after carotid endarterectomy. *Am. J. Pathol.*, **148**, 777.

53. Selby, P., Hobbs, S., Viner, C., *et al.* (1987) Tumour necrosis factor in man: clinical and biological observations. *Br. J. Cancer*, **56**, 803.

54. Galvani, D., Griffiths, S. D., and Cawley, J. C. (1988) Interferon for treatment: the dust settles. *Br. Med. J.*, **296**, 1554.

55. Feinberg, B., Kurzrock, R., Talpaz, M., Blick, M., Saks, S., and Gutterman, J. U. (1988) A phase 1 trial of intravenously-administered recombinant tumor necrosis factor alpha in cancer patients. *J. Clin. Oncol.*, **6**, 1328.

56. Loetscher, H., Steinmetz, M., and Lesslauer, W. (1991) Tumor necrosis factor: receptors and inhibitors. *Cancer Cells*, **3**, 221.

57. Perussia, B. (1991) Lymphokine-activated killer cells, natural killer cells and cytokines. *Curr. Opin. Immunol.*, **3**, 49.

58. Podlaski, F. J., Nanduri, V. B., Hulmes, J. D., Pan, Y.-C. E., Levin, W., Danho, W., *et al.* (1992) Molecular characterisation of interleukin 12. *Arch. Biochem. Biophys.*, **294**, 230.

59. Scott, P. (1993) IL-12: initiation for cell mediated immunity. *Science*, **260**, 496.

60. Lee, S. M., Suen, Y., Qian, J., Knoppel, E., and Cairo, M. S. (1998) The regulation and biological activity of interleukin 12. *Leuk. Lymph.*, **29**, 427.

61. Trinchieri, G. (1998) Proinflammatory and immunoregulatory functions of interleukin-12. *Int. Rev. Immunol.*, **16**, 365.

62. Brunda, M. J., Luistro, L., Warrier, R. R., Wright, R. B., Hubbard, B. R., Murphy, M., *et al.* (1993) Antitumor and antimetastatic activity of interleukin 12 against murine tumors. *J. Exp. Med.*, **178**, 1223.

63. Tannenbaum, C. S., Wicker, N., Armstrong, D., Tubbs, R., Finke, J., Bukowski, R. M., *et al.* (1996) Cytokine and chemokine expression in tumors of mice receiving systemic therapy with IL-12. *J. Immunol.*, **156**, 693.

64. Voest, E. E., Kenyon, B. M., O'Reilly, M. S., Truitt, G., D'Amato, R. J., and Folkman, J. (1995) Inhibition of angiogenesis *in vivo* by interleukin 12. *J. Natl. Cancer Inst.*, **87**, 581.

65. Tannenbaum, C. S., Tubbs, R., Armstrong, D., Finke, J. H., Bukowski, R. M., and Hamilton, T. A. (1998) The CXC chemokines IP-10 and Mig are necessary for IL-12-mediated regression of the mouse RENCA tumor. *J. Immunol.*, **161**, 927.

66. Atkins, M. B., Robertson, M. J., Gordon, M., *et al.* (1997) Phase I evaluation of intravenous recombinant human interleukin 12 in patients with advanced malignancies. *Clin. Cancer Res.*, **3**, 409.

67. Bajetta, E., Del Vecchio, M., Mortarini, R., *et al.* (1998) Pilot study of subcutaneous recombinant human interleukin 12 in metastatic melanoma. *Clin. Cancer Res.*, **4**, 75.

68. Motzer, R. J., Rakhit, A., Schwartz, L. H., Olencki, T., Malone, T. M., Sandstrom, K., *et al.* (1998) Phase I trial of subcutaneous recombinant human interleukin-12 in patients with advanced renal cell carcinoma. *Clin. Cancer Res.*, **4**, 1183.

69. Feldmann, M., Elliott, M. J., Woody, J. N., and Maini, R. N. (1997) Anti-tumor necrosis factor-α therapy of rheumatoid arthritis. *Adv. Immunol.*, **64**, 283.

70. Palli, D., Trallori, G., Saieva, C., Tarantino, O., Edili, E., D'Albasio, G., *et al.* (1998) General and cancer specific mortality of a population based cohort of patients with inflammatory bowel disease: the Florence Study. *Gut*, **42**, 175.

71. Correa, P. (1992) Human gastric carcinogenesis: a multistep and multifactorial process—First American Cancer Society Award Lecture on Cancer Epidemiology and Prevention. *Cancer Res.*, **52**, 6735.

72. Komori, A., Yatsunami, J., Suganuma, M., Okabe, S., Abe, S., Sakai, A., *et al.* (1993) Tumor necrosis factor acts as a tumor promoter in balb/3T3 cell transformation. *Cancer Res.*, **53**, 1982.

73. Komori, A., Suganuma, S., Okabe, S., Zou, X., Tius, M. A., and Fujiki, H. (1993) Canventol inhibits tumor promotion in CD-1 mouse skin through inhibition of tumor necrosis factor a release and of protein isoprenylation. *Cancer Res.*, **53**, 3462.

74. Robertson, F. M., Ross, M. S., Tober, K. L., Long, B. W., and Oberyszyn, T. M. (1996)

Inhibition of pro-inflammatory cytokine gene expression and papilloma growth during murine multistage carcinogenesis by pentoxifylline. *Carcinogenesis*, **17**, 1719.

75. Orosz, P., Echtenacher, B., Falk, W., Ruschoff, J., Weber, D., and Mannel, D. N. (1993) Enhancement of experimental metastasis by tumor necrosis factor *J. Exp. Med.*, **177**, 1391.

76. Malik, S. T. A., Naylor, S., East, N., Oliff, A., and Balkwill, F. R. (1990) Cells secreting tumour necrosis factor show enhanced metastasis in nude mice. *Eur. J. Cancer*, **26**, 1031.

77. Moore, R., Owens, D., Stamp, G. (1999) Tumour necrosis factor-α deficient mice are resistant to skin carcinogenesis. *Nature Med.* **5**, 828.

78. Orosz, P., Kruger, A., Hubbe, M., Ruschoff, J., Von Hoegen, P., and Mannel, D. N. (1995) Promotion of experimental liver metastasis by tumor necrosis factor. *Int. J. Cancer*, **16**, 867.

79. Anasagasti, M. J., Olaso, E., Calvo, F., Mendoza, L., Martin, J. J., Bidaurrazaga, J., *et al.* (1997) Interleukin 1-dependent and -independent mouse melanoma metastases. *J. Natl. Cancer Inst.*, **89**, 645.

80. Vidal-Vanaclocha, F., Amezaga, C., Asumendi, A., Kaplanski, G., and Dinarello, C. A. (1994) Interleukin-1 receptor blockade reduces the number and size of murine B16 melanoma hepatic metastases. *Cancer Res.*, **54**, 2667.

81. Naylor, M. S., Stamp, G. W. H., Foulkes, W. D., Eccles, D., and Balkwill, F. R. (1993) Tumor necrosis factor and its receptors in human ovarian cancer. *J. Clin. Invest.*, **91**, 2194.

82. Naylor, M. S., Stamp, G. W., Davies, B. D., and Balkwill, F. R. (1994) Expression and activity of MMPs and their regulators in ovarian cancer. *Int. J. Cancer*, **58**, 50.

83. Leber, T. M. and Balkwill, F. R. (1998) Regulation of monocyte MMP-9 production by TNF-α and a tumour-derived soluble factor (MMPSF). *Br. J. Cancer*, **78**, 724.

84. Negus, R. P. M., Stamp, G. W. H., Hadley, J., and Balkwill, F. R. (1997) Quantitative assessment of the leukocyte infiltrate in ovarian cancer and its relationship to the expression of C-C chemokines. *Am. J. Pathol.*, **150**, 1723.

85. Negus, R. P. M., Turner, L., Burke, F., and Balkwill, F. R. (1998) Hypoxia down-regulates MCP-1 expression: implications for macrophage distribution in tumors. *J. Leuk. Biol.*, **63**, 758.

86. Malik, S. T. A. A., Griffin, D. B., Fiers, W., and Balkwill, F. R. (1989) Paradoxical effects of tumour necrosis factor in experimental ovarian cancer. *Int. J. Cancer*, **44**, 918.

87. Gordon, H. M., Kucera, G., Salvo, R., and Boss, J. M. (1992) Tumor necrosis factor induces genes involved in inflammation, cellular and tissue repair and metabolism in murine fibroblasts. *J. Immunol.*, **148**, 4021.

87. Piguet, P. F., Grau, G. E., and Vassalili, P. (1990) Subcutaneous perfusion of tumor necrosis factor induces local proliferation of fibroblasts, capillaries and epidermal cells, or massive tissue necrosis. *Am. J. Pathol.*, **136**, 103.

89. Montrucchio, G., Lupia, E., Battaglia, E., Passerini, G., Bussolino, F., Emanuelli, G., *et al.* (1994) Tumor necrosis factor α-induced angiogenesis depends on in situ platelet-activating factor biosynthesis. *J. Exp. Med.*, **180**, 377.

90. Warzocha, K., Ribeiro, P., Bienvenu, J., Roy, P., Charlot, C., Rigal, D., *et al.* (1998) Genetic polymorphisms in the tumor necrosis factor locus influence non-Hodgkin's lymphoma outcome. *Blood*, **91**, 3574.

91. Stasi, R., Brunetti, M., Parma, A., Di Giulio, C., Terzoli, E., and Pagano, A. (1998) The prognostic value of soluble interleukin-6 receptor in patients with multiple myeloma. *Cancer*, **82**, 1860.

92. Chen, Z., Colon, I., Oritz, N., *et al.* (1998) Effects of interleukin-1α, interleukin-1 receptor antagonist, and neutralizing antibody on proinflammatory cytokine expression by human squamous cell carcinoma lines. *Cancer Res.*, **58**, 3668.

93. Woods, K. V., El-Naggar, A., Clayman, G. L., and Grimm, E. A. (1998) Variable expression

of cytokines in human head and neck squamous cell carcinoma cell lines and consistent expression in surgical specimens. *Cancer Res.*, **58**, 3132.

94. Tepper, R. I., Pattengale, P. K., and Leder, P. (1989) Murine interleukin-4 displays potent anti-tumor activity *in vivo*. *Cell*, **57**, 503.

95. Tepper, R. I., Coffman, R. L., and Leder, P. (1992) An eosinophil-dependent mechanism for the antitumor effect of interleukin-4. *Science*, **257**, 548.

96. Sun, W. H., Burkholder, J. K., Sun, J., *et al.* (1996) *In vivo* cytokine gene transfer by gene gun reduces tumor growth in mice. *Proc. Natl. Acad. Sci. USA*, **92**, 2889.

97. Forni, G., Giovarelli, M., Santoni, A., Modesti, A., and Forni, M. (1986) Tumour inhibition by interleukin-2 at the tumour/host interface. *Biochim. Biophys. Acta*, **865**, 307.

98. Golumbek, P. T., Azhari, R., Jaffee, E. M., Levitsky, H. I., Lazenby, A., Leong, K., *et al.* (1993) Controlled release, biodegradable cytokine depots: a new approach in cancer vaccine design. *Cancer Res.*, **53**, 5841.

99. Balkwill, F. R. (1994) Cytokine therapy of cancer. The importance of knowing the context. *Eur. Cytokine Netw.*, **5**, 379.

100. Owens, O. J., Stewart, C., and Leake, R. E. (1991) Growth factors in ovarian cancer. *Br. J. Cancer*, **64**, 1177.

101. Negus, R. P. M., Stamp, G. W., Naylor, M. S., Hadley, J., and Balkwill, F. R. (1997) A quantitative assessment of the leucocyte infiltrate in ovarian cancer and its relationship to the expression of c-c chemokines. *Am. J. Pathol.*, **150**, 1723.

Index